W9-AVJ-844

Final Analysis

Other books by Jeffrey Moussiaeff Masson

Against Therapy: Emotional Tyranny and the Myth
of Psychological Healing
A Dark Science: Women, Sexuality and Psychiatry
in the Nineteenth Century
The Complete Letters of Sigmund Freud to Wilhelm Fliess, 1887–1904
The Assault on Truth: Freud's Suppression of the Seduction Theory
The Oceanic Feeling: The Origins of Religious Sentiment
in Ancient India

Final Analysis

The Making and Unmaking of a Psychoanalyst

Jeffrey Moussaieff Masson

Addison-Wesley Publishing Company, Inc.
Reading, Massachusetts • Menlo Park, California
New York • Don Mills, Ontario • Wokingham, England
Amsterdam • Bonn • Sydney • Singapore
Tokyo • Madrid • San Juan

Library of Congress Cataloging-in-Publication Data

Masson, J. Moussaieff (Jeffrey Moussaieff), 1941–
  Final analysis : the making and unmaking of a psychoanalyst /
Jeffrey Moussaieff Masson.
    p. cm.
  ISBN 0-201-52368-X
  1. Antipsychiatry.   2. Masson, J. Moussaieff (Jeffrey Moussaieff),
1941– . 3. Psychoanalysts—United States—Biography.   I. Title.
RC437.5.M384   1990
616.89'17'092—dc20
[B]   90-38639
                                                                    CIP

Jacket design by Steve Snider
Text design by Wilson Graphics & Design (Kenneth J. Wilson)
Set in 10-point Palatino by NK Graphics, Keene, NH.

BCDEFGHIJ-MW-943210
Second printing, November 1990

For Daidie

Contents

Acknowledgments

My father and mother read through the manuscript of this book and approved of its publication. I wish to thank the following people who read the manuscript and offered various suggestions: Paula Caplan, Robert Goldman, Sally Sutherland, Terri Masson, John Friedberg, Tim Beneke, Bess Davoran, and Daidie Donnelley, to whom the book is dedicated. As with all the books I have written since 1980, Marianne Loring has been my constant helper, offering me her critical judgment and her splendid levelheadedness.

I especially want to acknowledge the collaboration of my editor, Nancy Miller. She was the editor for my first book, *The Assault on Truth*, and was practically a coauthor of both that book and the present one. She has read every word over and over, making innumerable suggestions, always with intelligence, tact, and good sense. I have been immensely fortunate to have her as an editor and friend.

I wish also to thank Jane Isay, who showed considerable courage in publishing this book and has been a source of encouragement and inspiration.

Final Analysis

Preface

In 1970, more than anything else I wanted to become a psychoanalyst. For the next eight years, as I underwent analytic training, I talked, read, and breathed psychoanalysis.

Final Analysis is the story of my psychoanalytic training; of my friendships with Anna Freud, Kurt Eissler and others, and of why I eventually left that world. If this were just my personal story, I doubt it would be worth telling. But the injustices and corruption I encountered as a psychoanalytic candidate were built into the very training process; the implications, I believe, reach out to every analysis and beyond that to psychotherapy in general. I was not the only person to notice these abuses, but by the time training was over, few of us had escaped the acculturation that made them, from the safe distance of graduation, seem necessary or even normal.

No book has yet told what it is like to undergo training as an orthodox Freudian psychoanalyst. Nor does any book tell what it is like to leave that profitable and prestigious profession—those who have been part of the inner circle of psychoanalysis either do not leave, or have left in discreet silence. Thus, until now it has been almost impossible to get an internal view of the workings of this "men's club" with its initiation rites; expectations of membership loyalty over truth; pressures to accept concepts handed down by the leader, no matter how irrational; xenophobic banding together against outsiders; and the punishment of anyone who poses questions or finally wants out. It is worth asking why no book like this has appeared

1

before, since people have written accounts of leaving almost every other cult. Why has psychoanalysis cast such an impervious net around those who become attracted to it and then train in it?

I cannot say that I undertook this book without reluctance. I hesitated for three reasons. First, there was my personal pain in reviewing events that were hurtful to me when they happened. It is sometimes alarming for me to look back at my own ignorance, naivete, or sheer blindness. I have winced many times in recollecting things I said and did during this period. George Orwell once said that nobody could ever write a truthful autobiography because the truth would always be too unbearably humiliating. Much of what I have remembered and written about here wounds my sense of who I would like to be.

My second reason for hesitating was that while I might be able to justify exposing myself to the ridicule, contempt, or mere head shaking of an anonymous public, I was not certain that I had the right to expose others who were close to me. Yet there seemed no way for me to write truthfully without doing so. All of the people I have in mind (for example, my mother, my father, my ex-wife) have read the book in manuscript, and while they were not in each instance delighted to find themselves in print in this manner, nobody objected or asked to have passages deleted.

Finally, I was daunted by the problem of accuracy. As an academic I had learned to rely on documents, and felt secure in their use. But now I was forced to fall back on my own memory. How accurate was it? Who can recall, exactly, conversations that took place years in the past? Fortunately, for some of the conversations in this book, including sessions with my analyst, I have notes made immediately afterward. But where I have had to rely on my memory and reconstruct conversations, I have tried to ensure that the words I ascribe to others are authentic by searching out original articles written by those I quote. Whenever possible, I have matched or confirmed the quotations with actual, published material. While I am convinced that the dialogues are as true to the spirit of the actual conversations as memory permits, I cannot suggest that they should be regarded as verbatim.

I am entirely opposed to fictitious elements in historical accounts. Nobody in this book was invented, in whole or in part, and nobody represents some aspect of another person. There are no composite characters here. But of course the names of my patients were changed, as were those of people whose names did not seem to me to add anything significant to the reader's understanding of my story or my thesis. I have alerted the reader to all fictitious names used by placing an asterisk after the name when it first appears.

Although this book stands by itself intellectually, it can be seen as the last volume of a trilogy that began in 1984 with the publication of *The Assault on Truth: Freud's Suppression of the Seduction Theory*, and continued in 1988 with *Against Therapy: Emotional Tyranny and the Myth of Psychological Healing*. *The Assault* was concerned with a specific problem in the history of psychoanalysis, but the question of covering up the truth led me to much broader questions. At the time I wrote it, I was still considering psychotherapy as a profession. If my criticisms of psychoanalysis were correct, I had to reconsider the ethical validity of practicing psychoanalysis. But was any other form of therapy more ethical? My next book, published in 1986, *A Dark Science: Women, Sexuality and Psychiatry in the Nineteenth Century*, made clear that the corruption I had observed in psychoanalysis was endemic to psychiatry's roots in Germany and France. Was modern psychotherapy, in any form, any better? Had it truly broken with its violent past? These questions led me to *Against Therapy*, with its historically grounded objections to the very idea of psychotherapy. While I had met kind and humane people who called themselves psychotherapists, none of them attributed their skill in helping people to their professional training; it was almost as if they were benign in spite of their training, and not as a result of it.

These books all resulted from reflecting on documents I

had come across during my brief tenure as project director at the Sigmund Freud Archives. But in none of these publications (nor in my edition and translation of the Freud/Fliess letters) did I give any of the personal background and details that led me to take the positions I did.

Throughout this book I acknowledge that Freud's preoccupations, with dreams, with memory, with the primacy of the emotions, with the importance of childhood and especially with human misery, are now our preoccupations, for the better. I am convinced that throughout history they have always been the preoccupation of men and women concerned with the betterment of their lives and those of their fellow creatures. We need to acknowledge Freud's achievements; we do not need to revere his errors. Freud was an extraordinary human being with all the failings of a man; turning him into an idol is a disservice to what must remain a continual search for truth. He taught us much; there is still much to learn.

But while I admire much of what Freud taught us, I do not admire the fact that he turned astute observations about human nature into elements of a vast and profitable profession with all the trappings of a jealously protected guild. The price for joining this fraternity is silence about its membership policy. Corruption is incorporated, not exposed; prejudice and bias have been accepted, even embraced. It is a high price to pay for membership. This book examines that fraternity and that price, and in the end describes a pathway to freedom.

Chapter One

ON BECOMING A PSYCHOANALYST

Toronto, 1970

"Dr. Masson, I'm glad to meet you. My name is Brian Bergman,* and I am a training and supervising psychoanalyst with the Toronto Psychoanalytic Institute. My job today is to interview you and then to give my opinion as to whether you are suitable for training to become a psychoanalyst. I have a number of questions to ask you. Please answer as frankly as possible."

Dr. Bergman leaned back in his chair and opened the interview:

"Have you been faithful to your wife?"

I was taken aback. This was hardly a typical question to be asked in an interview. I was being invited to reflect about what was, in my case, a difficult personal situation. How frank was it possible to be? And what were the consequences of being frank? I knew about free association, the only rule of psychoanalysis (at least the only overt rule), where the patient is told to reveal everything that comes to mind, no matter how embarrassing or how trivial or how outrageous. Here I was being urged, by the very nature of the interview, to engage in something like free association. But did I dare say everything that was on my mind?

One of the basic conditions of the much vaunted "analytic

---

*An asterisk after the first occurrence of a name indicates that a fictitious name has been used.

space" is that nothing said there goes beyond those walls. Confidentiality is at the very heart of all psychotherapy. Yet this was also an interview, and I knew that my answers were to be conveyed to others, who would then go on to make a judgment about my suitability for analytic training.

I swallowed hard. There could be no good answer. Many men, especially those who are themselves unfaithful, are convinced that most other men are too. So, if my answer were an unqualified "Yes, I am faithful to my wife," perhaps I would be seen as a liar, i.e., unfit to be trained as an analyst. "No" meant you were an admitted cheat and hence certainly not suitable for the highly "ethical" profession of psychoanalysis. I had only been married a short time, and in fact was already struggling with the issue of monogamy (I preferred this word to the more treacherous "faithfulness," which seemed to imply more than sexual monogamy). That seemed the more honest answer. I gave it. But what if now he pried further, and wanted the details of my struggle? After all, "I'm struggling" is pretty vague. Did he have the right? Was he like a customs officer, who could slowly and deliberately open your wallet and look at the pictures, receipts, credit cards, letters, anything he wanted? But Dr. Bergman moved on.

"Why do you want to become a psychoanalyst?"

I looked about the handsome room: good taste was in evidence everywhere. Contemporary drawings, plush leather armchair and couch, a small library. The building was situated on one of Toronto's busiest streets. Yet there was in this office a sense of quiet, as if we had entered a more leisurely century. It was December, snowing and very cold outside. In here we were warm. I could almost believe that I was about to be served tea and cakes. Dr. Bergman was a striking man in his early fifties, dressed comfortably in a corduroy suit with leather patches on the elbows. He was clearly Jewish, with a wrinkled, sad face. The eyes seemed to say, "I have seen it all. You can't shock me, you can't even annoy me." Obviously he was wise, accepting, warm, benevolent, and compassionate. The answer to Dr. Bergman's question was staring me in the face. I wanted to become a psychoanalyst to have everything he had. That struck me, though, as a bit too candid. What could I say, again, that would be honest yet not too honest?

"I am fascinated with Freud's work. No reading has ever absorbed me more. I could not think of a better profession than one that allowed me to read Freud as part of my training."

Dr. Bergman winced almost imperceptibly, and it suddenly occurred to me he might not have read Freud in many years.

"I can understand your wanting to study Freud, but what makes you think you would be good as a therapist, a practicing psychoanalyst? Why do you believe you can actually help anybody?"

That caught me short. I had almost forgotten that psychoanalysis was not just a theoretical discipline, but a clinical one as well. Psychoanalysis was not just something you studied but also something you *practiced*. I had no demonstrable talent in this area. I very much wanted to help, and hoped I would gain whatever skills could be taught, but there was no evidence that I had any particular talent for "helping" people. Again I thought I should sidle up to the truth, but not too close. "I am curious about other people's lives, and I'm bored silly with teaching Sanskrit at the University of Toronto. I want to change professions."

That was true enough. I had arrived in Toronto in July of 1970 and I was now an assistant professor of Sanskrit in the department of East Asian Studies at the university. I was twenty-nine years old, and this was my first real job. I had come from India the year before, where I had for two years done research on my Ph.D. dissertation for Harvard about the nature of suggestion in Sanskrit poetry. The main theories were written in Kashmir in the ninth century A.D., and I had translated the two key texts from the Sanskrit of Anandavardhana and

Abhinavagupta. I was no longer sure that I cared, however, about what somebody in the ninth century had said about the heart of poetry.

I had been lonely at Harvard. My relationships with others didn't seem to go deep enough to give me the sense that I was making permanent friends and becoming part of a larger community. I was unable to fall in love. I could easily imagine disappearing without leaving any trace in the world. This thought had a curious effect on me: it depressed me and yet the depression itself was so interesting a state for me to be able to feel, that I was nearly elated at experiencing it. But perhaps I am romanticizing my loneliness in retrospect. I know at the time that I just wanted it to end.

One autumn evening in 1962 I was walking in a quiet residential neighborhood of Cambridge, looking for the home of a friend. I stopped to ask directions at a house that looked cheerful and bright. The man who opened the door to me asked me in. He called his wife, and the three of us began a lively conversation, lively because both of them seemed to be unaccountably curious about me, where I had come from, what languages I grew up speaking, how I liked studying Sanskrit, where I was going that night, and whom I was going to meet and why. Both the man and his wife, it turned out, were psychoanalysts, the first I had ever met. I immediately assumed that their intense human curiosity must be a by-product of psychoanalysis, and I was fascinated. "What a wonderful profession," I thought, "that encourages such kindly intimacy." When I told them what I was thinking, and how badly I longed for just such conversations, they suggested I might be interested in therapy. They told me that Eric Erikson was involved in research for a new book on India and I might talk to him. He would probably understand my background and be able to refer me to someone who would be good for me.

I called Erikson, who was then teaching at Harvard, and he suggested I see another psychoanalyst. I am not sure what I expected Dr. Robert Zinmann* to be able to do for me in therapy, but I knew that the idea of somebody saying "Tell me everything" and meaning it was an unbearably exciting, heady thing for me. That somebody would first allow me to say everything that was in my mind, and then would *understand* it, promised a kind of intellectual and emotional utopia. It was the connection with another human soul that I was after.

I can't say that this therapy, once a week for about a year, "cured" me, but it certainly did not harm me. Yet even at this early date, some things that went on in my sessions unnerved me. I can remember entering the office for the very first time and telling Dr. Zinmann, "My problem is that I'm not able to fall in love, although I sleep with many women." Zinmann's response was a joke: "*I* should have such a problem." But it was a joke in bad taste, and its effect was not helpful to therapy. I would have a hard time, after such a comment, taking what this therapist said entirely seriously. Even as a joke, the statement revealed too much about Zinmann, things I did not need to know.

A few weeks later, I mentioned to a woman at a party that I was seeing Zinmann in therapy. She told me that she had been out with him, and had found him to be a "drip." Needless to say, I repeated this information in my session. Unaccountably, Zinmann could not remember meeting the woman. He pressed me for more details, and looked rather embarrassed, until I realized that the comment (understandably) had gotten to him. I sensed his vulnerability, and felt bad that I had exposed it, but I also felt somewhat deserted. I had wanted somebody stronger. It is not very surprising that at twenty I was looking for a strong father figure to help. But at the time I was surprised and sensed early on something of the power of "transference." It is rarely possible, when a human being is in deep need, to look upon somebody who offers help as merely another flawed human being with whom one is going to engage in a protracted conversation. A kind of wild idealization sets in, and we imagine the person in whom we confide to possess ineffable and valuable traits beyond those attainable by ordinary mortals. We ascribe value, and we project qualities onto this person that almost never correspond with reality. It is a little bit like falling in love— powerful emotions are called forth. It takes a strong person not to exploit the ensuing power imbalance.

Within a few months of arriving in Toronto in 1970, I received tenure at the university and was promoted to associate professor. My career was settled, and I could take the time to look at the world I had now entered. The narrow range of interests of my colleagues, the sheer dreariness of life as a junior academic, suddenly came home to me. I longed to escape the removed and impersonal world of academic scholarship into something more immediate and present. I wanted to live in the contemporary world. I wanted to know how people lived, what they dreamed about, what their marriages were like.

I had attempted to do something about my unhappiness at the purely personal level by going into therapy with a well-known Toronto psychiatrist. He was a man of broad interests, and was a lot of fun. Too much fun. He apparently derived great pleasure from our meetings, and this very pleasure made me feel my complaints were irrelevant and a sign of ingratitude. My transference took the form of wanting to be part of his life, or what I imagined his life to be like: filled with fascinating colleagues, a professionally successful wife, a lively family, a beautiful home in an exclusive area of Toronto, friends from the literary scene, and, above all, "wisdom" about human nature. So great was my longing that he seemed unable to resist it. Soon I was invited to his house and eventually became friendly with his whole family. That was the end of our therapy.

He had told me, honestly, that he could not help me in any event. My "problems" were deeper than what psychiatry was equipped to deal with. I needed psychoanalysis. I shivered with pleasure at the very word.

Soon after quitting that therapy, I was asked to participate in a television program about Indian poetry. I was intrigued by the woman producing and directing the show, Terri Thompson, a vivacious, attractive woman in her early thirties. She had a distinct accent, but one hard to place, and at our first meeting, I asked her where she was from. "Poland," she replied. I asked her if she knew anything about the Warsaw Ghetto, a subject that had long interested me.

"Yes," she replied, "I lived there during the war."

"That must have been hard."

"Yes," she answered, "that is why I am in psycho-analysis."

I was eager to talk with her some more. I asked her out to dinner, and she accepted. A few weeks later I was seeing her every day, and a few months later we were living together. In August 1971 we were married.

Terri and I were both intrigued by psychoanalysis. She had known, and I now became friends with, Michael Allen,* a professor in one of the departments of the humanities at the University of Toronto who was also in psychoanalytic training. It was unusual for an academic to be accepted for analytic training, especially if the field was not one adjacent to psychiatry (e.g., social work or clinical psychology). But when I talked to Michael about my interest in psychoanalysis, he said he could see no reason why I should not apply for analytic training, in spite of the fact that Sanskrit was about as removed from psychiatry as one could get. It was unusual, certainly, but it was not entirely impossible. I decided to apply.

Now that I was married, I was determined—for my sake, for Terri's sake, and for the sake of our future child—to rid myself of what seemed like an almost congenital sadness. I was convinced that psychoanalysis would have the answers to my personal unhappiness. Psychoanalysis, unlike psychiatry, did not interfere, it did not *intervene*, it attempted to understand. All an analyst did was sit and listen to you for fifty minutes, every day, five days a week, for months and even years, until your own memories coupled with his skills in understanding and interpreting yielded up the lost fragments, the missing pieces to the puzzle that was your personality. Total memory recovery was total cure. To unravel the past would, I thought, automatically lead to change in the present. The "aha" of insight was all that was necessary, I believed. The skills I would have to learn, then, were the skills of understanding. I believed that I would acquire these skills primarily through a process of unlearning; the ways I had learned to avoid my own feelings, to deny my own inner life, to evade personal conflicts, all these would be unlearned. To paraphrase Rilke, I was sure that psychoanalysis would be the pickax that would free the seas frozen within.

Outside, Toronto's searing wind was blowing hard, and the city looked forbidding. Never mind. I was here to unlearn, to be free, to bask in the warmth of Dr. Bergman's subtle understanding, kind humor, gentle intelligence. I was grateful to Freud. I was grateful to the profession he began. I was grateful to the Toronto Psychoanalytic Institute. I was grateful to Dr. Bergman. A whole new world was about to open up for me; I felt nothing but gratitude.

I was so grateful I told all of this and more to Dr. Bergman. I elaborated. I enthused. I gesticulated. When I finished, he looked slightly taken back, rose from his chair to indicate that the interview was over, but didn't say anything. How I wished, then, that free association worked both ways. I was disappointed that I learned nothing of what Dr. Bergman thought of my views. Did he find them naive, indulgent, romantic, or just wrong? Was he, as a psychiatrist, convinced I had misunderstood psychiatry? What would he tell his colleagues about me? In an ordinary job interview, you do not feel that you have just bared your soul. You get the job or you don't, but in this case Dr. Bergman was apparently going to decide whether or not I was worthy of becoming an analyst by intuitively judging my character and the quality of my entire life.

The interviews were an enigma to me. Indeed, the whole process of applying to the analytic institute was one shrouded in mystery. Before you could formally apply, you had to have been in analysis, what was called a "therapeutic" analysis as opposed to a "training" analysis, for some time. This therapeutic analysis was for you, not for the institute. Presumably it was required in order to demonstrate that you considered yourself to have as many foibles, not to speak of more serious illnesses, as did anybody else. I knew enough about psychoanalysis to know that anybody who believed he or she was free from "neurosis" was considered to be either psychotic or lost in irremediable denial, a bad candidate for analytic training.

The prospective candidate, who is almost invariably a psychiatrist, that is, an M.D. who has already finished a psychiatric residency, then applies to the institute. The personal analysis and the elaborate written application for admission (you are asked to write your first case history—about yourself) are supplemented by interviews with three senior training analysts. During the course of this first year "trial analysis," as it were, if you are accepted, you receive a letter telling you that the decision concerning your admission will be deferred for one year. You are urged to seek a personal analysis with a training analyst, an analyst who is permitted to train candidates. If your analyst reports, at the end of the year, that you are suitable material for becoming an analyst, you will be accepted. (How it is possible to associate freely, given this situation, was never explained, since the motivation to be "acceptable" is bound to influence what anybody is able or willing to reveal.)

This information about the application process, which has taken me only a few lines to relate, took many years to acquire.

In fact, when I received the letter saying that the decision concerning my admission had been postponed, I was under the impression that I had been rejected. There was no explanation. And almost everyone I approached was reluctant to give information and explanations. Knowledge of the process of psychoanalysis is a barrier to understanding, to the seeking of insight, I was told. Too much knowledge, in psychoanalysis, especially knowledge of psychoanalysis, is considered a bad thing. Many analysts recommend that you not read anything about the subject while you are in analysis. This is supposedly for your own good, so that you will not develop "intellectual" resistances.

The training analyst's report, at the end of one year, was made either formally or informally, indicating whether he believed you were suitable for candidacy. We were not told the details of the reporting procedure. I imagine that the theory behind this silence was that the anxiety would be insurmountable if we thought our analyst was going to make a report on us at the end of a year. Indeed, some analysts feel, correctly I believe, that "reporting" at any time during the personal analysis is a betrayal of the principles of analysis. In Toronto at the time, any training analyst could choose whether he would be a reporting analyst or not. But nobody, as far as I know, chose to tell the candidate which he was.

Dr. Bergman had been my first interviewer following my application to the institute. The second interview was equally enigmatic. The office was even more plush. But the analyst, Dr. Steven Garbin,* was austere. So austere, in fact, that he did not even greet me. He waited. And I waited. Finally, he shrugged— a signal, I took it, for me to begin free-associating about why I was applying to a training institute to become a psychoanalyst.

I went on and on. I knew I was revealing a great deal about myself, and I was certain that Dr. Garbin took either a strong liking to me or a strong dislike. Indifference seemed to me out of the question. I failed to realize that this may have been nothing but a mundane routine for him. He spoke very little, but at one point he told me that he had the impression that unusual "diseases" of the mind fascinated me. He mentioned Gilles de la Tourette syndrome or some other equally obscure psychiatric disease, and I eagerly nodded assent. Actually, I had no idea what he was talking about, but since he seemed to derive pleasure from having caught me out in some minor and amusing defect, I did not want to appear disdainful of his psychological insight. In fact, I believe this meeting planted the seeds of Dr. Garbin's later dislike for me. I think I just didn't appeal to him. Fair enough. We all have such feelings, all of the time. What is not fair is to believe that any person is above such feelings. Psychoanalysis is as helpless here as the single admonition "don't be prejudiced" would be. The most we can hope for is to be made more conscious of our biases so that we don't necessarily act on them in every instance. To be fair to Dr. Garbin, I think this was the case here, for I was admitted to the program.

Perhaps I was admitted on the strength of the final interview. Dr. Peter Thompson, an English analyst, was older than the other two. His room was rather shabby, certainly in comparison to theirs. He was pale, and in fact looked unwell. He was wrapped in several scarves, as if he had been cold for a long time. He spoke slowly. He asked me to recount any dream I had in the last few months that I considered significant. I told him the following dream: I was in Harvard Yard, listening to a brilliant woman speak to a small group of students that had gathered around her. The woman was animated, and I thought her to be extremely attractive, though in fact in the dream I was aware that others probably did not. She had a prominent, aquiline nose that I found particularly beautiful. I was lost in admiration and thought to myself how much I wished I knew the person, when, to my astonishment, she made reference to me, and then spoke directly to me. I suddenly realized that in fact I was with her. I was flooded with a feeling of great happiness.

Dr. Thompson asked me what the dream made me think

about. Of course it made me think about my wife, whom I had only recently married. Terri was from Poland, and had a very striking "Jewish" nose. Many people found her attractive, others did not. I was deeply taken with her extraordinary intellectual abilities. In "real" life I was lamenting my continued attraction to conventionally beautiful women, and I mentioned this to Dr. Thompson. He surprised me by letting down his analytic guard—his objectivity—telling me in a voice filled with genuine sympathy and pleasure that in his opinion this dream was very important, it was telling me something significant, I should listen to it, and think about it, and I would, over the years come to see it as critical to my life. I was astonished by what I took to be almost a prophecy. After all, an analyst like Dr. Thompson was not supposed to function as a fortune-teller. How could he know, from my telling him a simple dream I had the night before, that it would loom large in my life? I was torn between the skepticism I felt for such necromancy and gratitude for the kindliness of the sentiments he was expressing. While I'm not sure that the dream did prove to be critical in my life, what he told me was benevolent, and it was, I am certain, well-meant advice, though the attempt to clothe it in the mysterious omniscience of psychoanalytic language was unnecessary. Moreover, he was telling me something personal, something *he* believed and felt, and I was grateful to him. The kindly instinct he manifested that day permeated the interview with him, and made me realize that surely here was a quality no analyst could afford to be without. Yet neither Dr. Garbin nor Dr. Bergman showed it. Indeed, I did not see how compassion, kindliness, and sympathy could be artificially conjured up.

I worried for myself. Did I have a good heart? Yes, on balance, I thought I did. But what Dr. Thompson had just shown me was not merely a good heart, it was an *open* heart, one that was not afraid of emotional generosity. Did I have that as well? On balance, I would say no, I did not. Was this a quality I would gain, as part of my training? I did not see how I could acquire an "open heart" merely by reading, or by studying. Perhaps, though, it would come in my own analysis. Or perhaps it would come in the course of treating patients. Maybe when I was directly exposed to other people's suffering, and I was in the position of being expected to help them, maybe, just maybe, I would rise to the occasion. But I must admit that my doubts were strong and growing.

Chapter Two

THE WORM OF ANALYSIS

Psychoanalytic training is three-pronged: the personal analysis, considered the most essential aspect of training; the seminars, with which nobody was satisfied (too long, too frequent, too late in the evening); and finally, our own cases (before graduating, every candidate had to analyze at least three patients, a kind of practice analysis under supervision). The personal analysis continued throughout, taking a minimum of four to five years, and frequently in fact as many as ten years. We began taking the seminars immediately after admission, but we would see patients in analysis only when the committee decided we were ready, generally two to three years after beginning our own analysis, and a year or two after we started taking the seminars. We would not be admitted to the so-called scientific meetings, meetings where cases are discussed by the senior analysts, until after we had begun seeing our second case. Generally, in Toronto, the training lasts for six to eight years, at which time we would be "graduated": we would become members of the International Psychoanalytical Association and could call ourselves psychoanalysts.

On September 12, 1971, I waited for Dr. Irvine Schiffer to open the door to his analytic office, located in an apartment building in the northern part of Toronto. I was there for my very first session, but this was the place I would be coming to for the next five years, five days a week, fifty minutes a day.

As part of my training, I was required to undergo analysis, whether I wanted to or not. My behavior, whether in the external world, as revealed by what I reported back to my analyst, or in my internal world as revealed by my wildest dreams, my most shameful fantasies, my most secret thoughts, were to be laid out before this man with nothing held back. What he chose to do with these revelations was something I could not know about. He might choose to report, or he might not. He might talk about me to the other training analysts, or he might not. Certainly if he thought, at whatever point in the analysis, that I was not worthy of becoming an analyst, then he had an obligation to my future patients to report this opinion to the training committee. Had not Freud demanded that Paul Federn reveal details of his analysis of Wilhelm Reich, in order to provide Freud with the ammunition he needed to rid the movement of this man?

The whole point of the personal analysis was for the candidate to learn enough about his own problems so that they would not interfere with his ability to remain objective when faced with patients of his own. The analytic space was like an operating theater. It had to be kept as pure and unsullied as possible. The influence of the outside had to be kept to a minimum. The analytic instrument must be immaculate. Only a personal analysis could ensure that the analyst would be free of personal bias and prejudice. This personal analysis would last as long as necessary. Often called a "training analysis" when one is a candidate because it is part of the training program— it is the psychoanalysis you undergo at the hands of an experienced analyst both to solve your own problems and to show you how analysis works. It was no longer called a "therapeutic" analysis once you were accepted by the institute as a candidate; the word "therapeutic" was then reserved for "real" patients who were in analysis only for their own good. There were rumors that some of the senior training analysts, so called because they and only they were licensed to analyze candidates, were themselves, twenty or more years later, either *still* in analysis, or constantly going back for refueling.

In any event, the personal or "training" analysis would not last less than three or four years. An average analysis was five or six years. At $75 to $100 an hour, five days a week, the expense of being trained was considerable, even if the government did pick up part of the tab (in Canada, medical insurance, which is universal, pays for at least 60 percent of an analysis, whether personal or training). I was paying $75 an hour, while earning a salary at the university (at least initially) of less than $20,000 a year.

A training analysis differs from what is called a therapeutic analysis in that you can fail a training analysis, but you cannot fail a therapeutic analysis unless you fail to pay. Some people feel that this training analysis is therefore not a "real" analysis, and I have heard analysts say, years after their training was over, that they wanted to go back into analysis because "the training analysis was for the institute, this one is for me."

What did I know about Dr. Schiffer? Nothing much. He was a psychoanalyst. He was the president of the Toronto Psychoanalytic Institute. He was a psychiatrist. He had been a neurologist. He had been trained in Boston. He was in his sixties. He was Jewish. His own analyst had been Felix Deutsch, who had been, for a time anyway, Freud's personal physician. (Deutsch fell out of favor when he concealed from Freud that he was ill with the cancer that was to kill him. Freud roared when he discovered it, *"Mit welchem Recht?* [With what right?]," and Deutsch was banished from the inner circle.)

I had first heard of Dr. Schiffer from the psychiatrist I was seeing in therapy, who had told me, "There is only one analyst who could handle you: Irvine Schiffer." These were magic words for me; it was as if I really didn't need to know anything more about him. Moreover, my therapist had called him "an analyst's analyst," which I took to be high praise (only later did

I learn that it is a quasi-technical term referring to an analyst who trains candidates). "He's the best," he added, and it did not occur to me then that this was probably only a cliché. Later I learned that my therapist knew almost nothing about Schiffer, having spoken with him once for a few minutes. My mother, like many other Jewish mothers, had often used the phrase "the best." You could not see a doctor unless he was "the best." When I mentioned him to Mike Allen, who by then had become my closest friend, he told me that in fact he had been in analysis with Schiffer himself, and gave me to understand that Schiffer was, indeed, a remarkable analyst. I don't, to this day, know what I could have done on my own to find out more about the man to whom I was about to entrust my emotional life, but I do know now that trying to find things out about one's training analyst seemed to be actively discouraged. To do so is almost considered a kind of lèse-majesté.

Dr. Schiffer opened the door and called me in. He was a man of medium height and slim build with a sallow complexion and, of course, piercing eyes. I was surprised by his office. I had been in the offices of several analysts before, and all had been, to my mind, in good taste. This one was not. It was sparsely furnished, what furniture there was ranging from gold ornate to Woolworth schlock. The couch was black vinyl with a crumpled Kleenex on the plastic pillow. There was a large, empty desk, with a black Naugahyde chair. Dr. Schiffer was smoking a cigar. Was I meant to see Freud? The room smelled of this and of a lack of ventilation. I thought, Am I really going to spend the next five years in this room?

Dr. Schiffer asked me to sit down. There was some minor business to conduct. The ground rules of my analysis were simple: I was to come every day at twelve-thirty; sessions were for fifty minutes; I would pay for any sessions I missed; the charge was seventy-five dollars per session; I would be given a bill every month, and was expected to pay at the next session; I would come for the full year, with the exception of two weeks in August; if I wished to take my vacation at any other time, I would have to pay for the sessions.

Apart from these external matters, there was only one basic rule: free association. I was to say *everything* that came to my mind, no matter how embarrassing, how unpleasant, how rude or vulgar or silly. I was to hold nothing back. Also, I would be well advised not to make any major life changes while I was in analysis. Did I have any questions?

I had a million. I could not even begin. I was silent.

"Good. Please lie on the couch."

What a strange feeling to be lying on the couch for the first time. I could not see Dr. Schiffer, though I knew he was right behind me, smoking his cigar in his chair. It was such an incongruous situation that I could only wonder how I would ever feel comfortable with it. I felt as though I was being initiated into some secret rite. I did not like cults, rituals, secrets. But this one seemed so adult, that I mistook my legitimate discomfort for something else. (My disagreements would later be interpreted as "adolescent rebellion.") Why were we doing these odd things?

I lay there silent for a moment, thoughts swirling through my head. Suddenly free association seemed like the hardest thing in the world. How could I even begin to sort out the thoughts (don't sort them out) in my head, and tell them as they were occurring to me?

And what about the thoughts I had as I came in? I couldn't stand Schiffer's office. Did I dare to tell him that? Could I tell him I couldn't bear the cigar smoke? Could I tell him that already I was beginning to believe I had made a mistake in my choice of analyst?

Michael Allen, who had been in analysis with Schiffer, hadn't told me much about him, claiming it was best that way, for my analysis. But I knew he admired Schiffer. Should I not have investigated Dr. Schiffer on my own, however? Looked up his articles? Read his books? I tried, briefly, but could find nothing. He had written almost nothing. Was this a good sign or a bad sign? I took it to be positive: he *chose* not to write, so that he would reveal even less to inquisitive patients like me. What did I need to know anyway? That, I decided, was why his office was so sparse—he didn't want to give anything away. Was he married? Did he have children? I knew I would never know, unless I found out in a roundabout way. He would never tell me. Well, I really can't lie here the entire session, thinking my own thoughts when I have just been told that I have only one obligation and that is to say everything that comes to mind. I had better start with these last thoughts. "I was lying here thinking I would probably never know if you were married or had children."

"Wrong. I am married, and have four children."

Long silence. I was struggling with assimilating this and wondering just how free association worked. How was I supposed to put what I was thinking? Could I say, "How could you tell me such a thing? Isn't that against the analytic rules?" Could I say, "Wow! My friends were right, you really are different"? Did I say, "I'm a bit apprehensive"?

I said, "I thought you could not answer my questions."

"What? Nobody tells me what I can and cannot say in my office. This is my office. I do what I goddamned well please in here."

The thoughts were coming even faster now. This man was definitely different. At least things were not going to be dull. But I still didn't know how to keep up with my thoughts. It was like learning a whole new language. I had never been in a situation before where I could, no, where I must, say everything that occurred to me. Quite the contrary. I was constantly restraining myself, trying to remember to be courteous, to hold myself back, not to be rude, nor to reveal more than was asked for. It was not easy. I am impulsive. I like to talk. I like to interact. I like to become personal quickly and talk about and ask for details that are ordinarily not revealed until much later, if ever. Well, here was my chance. But I was still stymied. I liked, and had no problem with, talking about myself. But I had never tried to say everything that came to mind to another person. And here was another person asking me to do just that. Did he really mean it? Could he really take it? Would it, as I had been assured, have absolutely no effect on him, or what he thought of me? It seemed so unlikely. For example, his comment that he could damned well say what he pleased in his office made me somewhat apprehensive. It was not a typical "analytic" phrase. What did it mean? What did it presage? It sounded ominous. Like he was angry, or prepared to be angry easily. I was not sure I liked it. He was moving awfully quickly, without regard for social niceties, considering he knew nothing about me.

I wondered if indeed this was a special "space," a place set apart, where absolutely anything could be said. That was certainly the theory, and it held a certain attraction for me. Imagine a place out of time, out of bounds, where you could retire for a short while, and do or say anything you wanted, without its *counting* in the real world. Okay, I said to myself, here goes: "You know, I have to tell you that I was disappointed when I came into your office. I was expecting something more elegant, nicer, with better taste. I see a lot of tacky furniture here, and everything is functional and ugly."

I paused, expecting him to wait for me to continue. He didn't. "Listen—this is my office. I furnish it any way I please. You don't like it? Leave. As for the 'tacky' furniture —fuck you, my mother died recently, and those were her things. I am proud to have them here."

Whew. Touchy. Amazing, really. He was reacting to me, in his words, exactly as I reacted to him in my mind. *He* was free-associating. But wasn't that against the rules? Wasn't that my job, with him just listening and making interpretations from time to time? Wasn't he supposed to just nudge me in the direction of insight? Get me to where I was just a step away from seeing a connection, then point me gently in the right direction with a well-placed question, a nod, a rising tone?

This was altogether different. I had never been with any-body before who said whatever came to mind. I told him so, adding that it didn't seem to conform to normal analytic rules.

"Well, welcome to Irvine Schiffer's office. As for analytic rules, it's all a bunch of flea fucking as far as I'm concerned. Here, I make the rules. I do whatever I please. I say whatever I please." I could see that I was not going to get a chance to do anything but react to what he was saying, so many thoughts were crowding in on me.

"You mean, you will answer my questions?"

"If I feel like it."

"Okay, I'm going to test you. What do you think of Dr. X?"

"I think he's an anti-Semitic, pretentious fart. I can't stand the guy."

"Dr. Y?"

"A vapid twit—no guts. Wouldn't trust him as far as I could throw him."

"Dr. Z?"

"A little snake. No soul. You happy now? You get what you wanted? Are you ready to begin *your* analysis?"

I was speechless. No blank screen here. No mirror. How could he tell me these things? How did he know I would be discreet? Why should he trust me? Or is it that he didn't care what I said? I had to know. "Why are you telling me these things?"

"Look, I'm a free man. I think what I please, and I say what I think. I don't care who knows it."

But I wondered, what about its effect on me? I was not an acquaintance; I was a patient. Surely the way he talked with me, what he said to me, what was expected of our relationship, was different from what it would be like in the "real" world. Schiffer agreed. But he had a somewhat idiosyncratic view of the notion that the analytic space was a protected one, a unique environment. For him this meant that this place must be one of complete and total honesty. Not only did I have the obligation to say what I really felt or really believed, *he* had the same obligation, and was to be kept to the same strict standard. Truth, he proclaimed, was our only guide.

I found this exhilarating, at least theoretically. But I also wondered whether it was possible to achieve in practice. Were there to be no formal rules governing his behavior toward me? Could he really say anything he liked? Would he not be held back by what he believed would hurt me, humiliate me, damage me? I had, after all, handed myself over to his care. He was also, as he reminded me several times in early sessions, a physician, and so my welfare was important to him. I was not sure I believed these pious sentiments, but I was comforted by them. Like a surgeon (another analogy used by Freud and by Schiffer), he would use pain only to relieve pain. It might hurt for a while, but the purpose was to heal. I liked the sound of this analogy, and I liked the medical metaphor. I did not consider, then, the potential for abuse that such an idea has built into it. Then, I was only struck by the novelty of being in a room with a man I considered wiser than myself, who would tell me everything that occurred to him as I revealed my whole life to his scrutiny. But I was also puzzled, even alarmed by the power he held over my professional life. What went on between us was not *really* private. This could, conceivably, act as a forceful threat. On the face of it, there did not seem to be any possibility that he could now or later want anything from me. I was the supplicant. I was the "sick" one, the person needing help. I also needed his approval to become an analyst. It was a situation fraught with danger. I was not sure that Schiffer, or anybody else, could steer a way through it.

When I ask myself why people submit voluntarily to analysis, I know that part of the answer lies in the great pleasure there is in being able to talk about yourself, indeed, being mandated to talk about yourself, and only yourself, for almost a whole hour, every single day, for years, and to be assured of a sympathetic listener. (I did not stop, then, to ask how anybody could possibly know that the listener would be genuinely sympathetic.)

People would often ask me, "How could you find anything to talk about for a whole hour, over so many years? Didn't you get bored with yourself? Didn't you get tired of hearing the same things?" One thing I can never complain about in those sessions with Schiffer was boredom. He was not boring, and the procedure was not boring. I was not accustomed to telling my dreams to somebody, every day. This in itself was a fascinating thing to do. Already you are faced with a problem: How do you convey the complexity of a dream, and especially how do you put obscure dream-feelings into words? *Telling* a dream was a lot harder than I thought it would be. Conveying the atmosphere in the dream is even more difficult. Over and over I would find myself stopping, hesitating, and saying, "No, that's not it. I haven't gotten it." And, even worse, sometimes, right in the middle of my telling it, the dream would begin to fade. I would lose some sense of the dream, some feelings about it. The more I would try to convey how the dream felt to me, the more remote it became, like a picture fading in front of my eyes, or like a landscape that was rapidly receding into the background. Then this fact itself would become the subject of discussion—that is what free association is all about—and one could go on, I could anyway, for hours.

Not that I did. Not even minutes, really. Because I was up against an equally irresistible pressure on the part of Dr. Schiffer to free-associate back, to answer my thoughts with thoughts of his own. This doesn't sound so bad, you might think; two intelligent people, letting down their social guards, and saying whatever came into their minds, taking each other as their major topic of thought. But this is to forget something important: we were not two equally matched people. Ours was not a love affair, it was not even a close friendship—this was psychoanalysis. I was a patient. Dr. Schiffer was my psychoanalyst. There was no parity. Instead, there were "rules," even if he seemed to have his own version of many of these rules. In fact, I suspect that most practicing analysts have their own versions of analytic theory and analytic technique, and they do not invariably share them with their colleagues, or ever write about them. Their patients know about them from direct experience, and not always to their advantage.

Dr. Schiffer, for example, seemed to believe in sometimes allowing his bad temper full play in the analytic session. The first time it happened I was dazed. It was a simple matter. My appointment was for twelve-thirty. I would come a few minutes early, park my car in the back of the apartment building on the ground floor of which Schiffer had his office, come into the waiting room, pick up the one abominable journal he mysteriously kept there—*Medical Aspects of Human Sexuality*—and wait to be called. From the very first session, and routinely thereafter, he was late. Five, ten, fifteen, twenty, or even thirty minutes. For the first few months, he would pretty much keep to the fifty minutes by extending the sessions. But after a while, he became more erratic about time: sometimes he would tell me the session was over when we had been going less than half an hour; at other times, he would go for sixty minutes or more. I assumed he was keeping track, and that I was being given more or less an average of fifty minutes. After a few months, Dr. Schiffer was taking me later for about half of all my sessions. In fact, I would see his car pull up in the parking lot forty-five minutes after I got there. He would come in the front door, open the door to the waiting room and call me in.

I have always found it difficult to be kept waiting, and I began to get more and more angry. But for some reason, believing perhaps that in his eyes I would be stepping beyond the boundaries of the basic rule of free association, I was reluctant to bring this up. I thought that perhaps I could remedy the situation on my own by coming forty-five minutes later. I could just pretend the session was supposed to be at one-fifteen instead of twelve-thirty. I tried this. A few times, when Dr. Schiffer was also late, it worked out fine. But a few times he was waiting, and I found that he was counting the fifty-minute session from the original twelve-thirty appointment, so that in effect our time was only five minutes. I could always hear the next patient arrive and sit in the waiting room, and no doubt Schiffer could as well and would be under some pressure to stick to a time schedule. Sometimes this patient would be kept waiting, and sometimes I was dismissed early, but somebody always lost out, except Dr. Schiffer. Not that he was trying to cheat me, but his erratic arrivals seemed to indicate that he didn't feel my time was as important as his. Clearly I had to remark on it.

I did so on a day when he was even more than forty-five minutes late, and I could feel my blood begin to boil in anger. I knew that I had to talk about it. But instead of *being* angry, I spoke *about* my anger, and then simply told him the problem. He made the reasonable point that the time allotted to me was from twelve-thirty to one-twenty, and if I chose to use that time in other pursuits, that was my business, but I then had to pay for it.

"But the rule, I take it, does not apply to you, correct? You can come when you like?" There was an edge to my voice, but he chose to ignore it.

"Correct."

I felt I should still control any expression of my anger and explain my position. "But Dr. Schiffer, this puts me in something of a dilemma. Since you are late about half the time by as much as forty-five minutes, I have to come at the scheduled time, but often do not leave until after two P.M. This makes it difficult for me to make any appointments of my own on a regular basis, since I am never sure when I will be back in my office at the university. It also means that these sessions take up, with coming, going, and waiting, almost two and a half hours. Moreover, even more upsetting to me is that I don't always get my fifty-minute session. Now I know you don't care about 'time' to the extent I do, and I am sure you try to make up for lost time, but I get the distinct impression that I am not getting, all things considered, my full time from you."

"Too bad."

That was more than I could take. Finally there was real anger in my voice as I told him I was fed up, and he could bloody well be on time if I was expected to be on time. I had waited more than forty-five minutes today, and I was tired of it.

"You're a liar."

"I beg your pardon?"

"You heard me. You're a goddamned liar. You were forty-five minutes late. I was waiting for *you*."

Something was very wrong here. *He* was lying. I told him that. His anger rose. He did not dispute the time now, but he began to imitate my voice, mocking what I had said earlier about having appointments. "You have your appointments. Very important no doubt. *Dr.* Masson. *Professor* Masson. Don't make me laugh. You're just a candidate, remember that."

Well, maybe he had a point. Maybe this was his curious way of pointing out my grandiosity. But what about *his* grandiosity? Why should his appointments and his schedule and his time be so much more important than mine? And who, after all, actually knew whether they were or were not? And was he not saying this merely to protect himself, or worse, just to hurt me, because I had the temerity to attack him? I was surprised that I had even dared to get angry at him, but this interesting new development did not seem to interest him in the least. Suddenly we were not in a "metasituation," this *was* the situation. We had stopped talking *about* something, we were in the middle of doing it. But the "it" we were doing seemed to have more to do with Schiffer than it did with me. Yet these were my sessions. I was paying for them. And I was not paying him to learn about *his* vulnerabilities.

Why had it taken me so long to speak up? What was I afraid of? I was afraid of Schiffer the man, rather than Schiffer the analyst. It was becoming apparent to me that it would be very easy for us to step out of this so-called protected environment of the analysis out into the real world, where Schiffer could do me real harm. Analysts often claim that the analytic room is no more than a dream-space, and the analytic relationship but a metaphor. This is not true, however. What happened in this room did *not* leave me unaffected. He was *not* only there to understand me or help me to understand myself. He was a living creature of flesh and blood and there was nothing metaphorical or dreamlike about the anger he was now expressing directly at me. Nor did it strike me as a wild or paranoid thought that this anger could affect my career directly. He held genuine power over my professional life.

In the past, when I had made some oblique reference to time, Schiffer had boasted that he was not a slave to time, that he was not bound fast to society's silly rules about being on time. An analysis of the depths of a person's soul, he explained plausibly, could not be expected to watch a clock. I had read that Jacques Lacan, the French analyst, sometimes had sessions that lasted only a few minutes—what mattered was how quickly he reached the core. But how disinterested could he have been in time when he seemed to dismiss patients after only a short session, never saw them for an entire day, and always charged them the full fee? An analyst's freedom from the normal servitude often seemed to work exclusively to his own advantage.

What I was facing in Schiffer now was no longer a theoretical disagreement. He was breaking our basic agreement: a fifty-minute session beginning at twelve-thirty. Moreover, he had accused me of being a liar. And he had cursed me. This was strong stuff to be hearing from your analyst. I said so. How could I be expected to speak my mind in the future, if this was the reaction I got? Anger in the real world can be worked through mutually. But here I was the suppliant, needing both his help to become a better person, and his permission to become an analyst. He held all the power, and he knew it. These kinds of "insights" were intolerable to him; they were a waste of his time, he maintained, since they did not focus on me. And anyway, at this point he was no longer interested in reasoning. I could hear his voice rising, I could feel the tension, when he suddenly said in a kind of growl, "You're a goddamned fucking liar. You make me sick, you son of a bitch." This was no longer psychoanalysis. I could not lie on the couch and muse about what I had just heard. I sat up and turned around to face him. His veins were throbbing, he was red in the face. If this was an analytic ploy, a test, a trick of the trade, then Schiffer was the world's greatest actor. I told him I was leaving, and I walked out.

I could not know what had really provoked his outburst because I could not know what was happening to him. It was useless for me to speculate, since I had so little to go on. It was more valuable to bring it back to me. Perhaps Schiffer was trying to bring home to me one of my faults forcibly in this dramatic way, showing me directly, without the mediation of speech, something about my interactions with other people. Maybe this was, after all, just an analytic device, an unorthodox one to be sure, but perhaps such departures from the norm were necessary to crack such a hard nut as me. I calmed down.

But I also felt that I really didn't know enough to understand what was happening. I tried that same evening to talk about the session with my friend Michael Allen, who had been in analysis with Schiffer before me. He responded with some hesitation. I thought at the time that he did not want to interfere in my analysis by commenting on Schiffer's behavior, or mine; this was something I had to take up in the analysis. I knew that Allen and Schiffer were preparing some sort of university program in psychoanalysis together, and I could imagine that he did not want to compromise his relationship with his former analyst by venturing an opinion. He knew, of course, that I would have to talk about it the next day in my session, and apparently he was being careful.

As I write these words, I have just found in my files the original notes I wrote after a very similiar session.

Today is Monday, June 17, 1974, and I have kept this bottled up all day . . . Should I tell Terri? Schiffer has severely warned me never to tell her anything of what goes on in my analysis. A consultation? With whom? I once suggested it (the day he called me a filthy Nazi and cursed and shouted) and he wouldn't hear of it. He said I either stuck it out or I quit. [I don't remember the exact words, but] he implied I had no guts, no courage, no stomach for the truth. I don't know what will happen tomorrow. I want to sit up and tell him all this, but I fear he will throw me out. Whom would I go to then? Would it mean quitting the Institute? How could I ever tell someone else in Toronto all this? Especially a training analyst. I could perhaps speak to [Terri's analyst]; Terri thinks so highly of him. But I know he sends many people to [Dr. ———], and I can't respect him. Also, how would it affect Terri's analysis? And could I trust him not to tell? The person I really want to speak to about all this is Mike Allen. But I don't know if I can. I genuinely like and trust him. But will he be able to admit what's going on? Schiffer was his analyst. . . . he may have had similar experiences with Schiffer and may be able to explain. Of course he may refuse to listen when I tell him what it's about. And if he recognizes the truth, what can he tell me?

I decided that the best step would be for me to get an outside opinion, to speak to another training analyst about what was happening. The next day I opened my session by suggesting this to Schiffer. The effect was not what I expected: if I went outside of the analysis to settle this matter, our relationship was through. No other analyst had the right or the ability to judge him. And whom would I go to? We had already both agreed that they were for the most part a bunch of losers. Schiffer said I only wanted to go out of a desire for vengeance, and if I did, he would see to it that I lost the battle. He would claim I was paranoid, and delusional, and that I was inventing it all. He would calmly deny my story. He said, "And whom, Masson, do you think they will believe? You, a first-year candidate, not even a medical doctor, or me, Irvine Schiffer, training analyst, and an officer of the Toronto Psychoanalytic Institute?"

He was right, but again, what struck me was not the logic of what he was saying but the lack of analytic distance it implied.

He was threatening me, as if we were locked in some political battle. Moreover, the kind of politics he was playing with me was dirty. He knew very well that I was telling the truth, yet he was willing to admit to me that he would lie. To save his skin? I asked him.

"Not at all," he explained, now very calm. "I am doing it to save your analysis. You cannot get analyzed by anybody in this town except me. I'm your only analytic friend. You are your worst analytic enemy. I will employ any means necessary to save you. Lie, cheat, steal, it's all the same to me, as long as you get better. And you know why, Masson? I'm going to tell you. Because I think you have the potential to be not just a good analyst, but a great analyst. I think you can do important things in this field. Things that nobody else has yet done until now. You see, I have trust in you, I have confidence in you. Far more than you do in yourself. I do what I do out of love."

It was a powerful speech, and I have tried to reproduce it exactly as I remember it after all these years. It certainly galvanized me. It shamed me. How could I have been so callous? How could I have been so mean-spirited toward this man who had only my good at heart? Who had it to such an extent that he was willing to forgo so many of the rules of analytic procedure that he had learned, merely to save me from myself and for analysis? How could I be so small-minded as to believe he had a political agenda?

Schiffer had gotten me in my narcissistic weak spot. I *did* believe I was special, of course, like every other candidate. But the spot he touched in me went even deeper than narcissistic gratification. Since I was sometimes overly skeptical about the capacity of other people to engage in selfless acts, I now found myself inclined to believe that Schiffer was one of the few exceptions, and that I had just wanted to tear down this goodness by looking for the dark side. I was, explained Schiffer, merely projecting my own defects onto him. How much better it would be if I could learn from him, use him as my ego ideal, pattern myself after his greater maturity.

Such was his advice. Theoretically I did not find any of this particularly convincing. But I could accept his version of events, and stay in analysis, or I could quit. In fact, he told me this in so many words: "You can trust me, or you can quit. But don't think that any other analyst would take you. If you leave me, it will appear—in fact I'll say it—that you're no good, you can't be analyzed." He was putting my career on the line in no uncertain terms. When I pointed out that this was a threat, and an ugly one, Schiffer maintained that he was only doing it for my own good. He would, he insisted, do just about anything to help me, and helping me meant keeping me in analysis with him, since he knew, as I could not, that nobody else could help.

Part of the problem was that while the concept of an ego ideal toward which one could strive, of a benevolent identification with a wise father figure, did not appeal to me intellectually, it had enormous emotional resonance. I did not like what Freud wrote in 1937, toward the end of his life, in "Analysis Terminable and Interminable," but I could not deny its power: "It is therefore reasonable to expect of an analyst, as part of his qualifications, a considerable degree of mental normality and correctness. In addition, he must possess some kind of superiority, so that in certain analytic situations he can act as a model for his patient and in others as a teacher." However, Freud had gone on to add a wonderful sentence: "And finally we must not forget that the analytic relationship is based on a love of truth—that is, on a recognition of reality—and that it precludes any kind of sham or deceit." This, I thought, is what Schiffer has taken to heart. I still yearned then, and probably even more so earlier, for a strong, masculine person on whom I could pattern myself. Somebody I could admire, and imitate, and become close to and learn from. I had sought this, always, in my teachers, and my search had always ended in disappointment. I was attracted emotionally to a position that I could only despise intellectually. In 1914 Freud had already written in "On the History of the Psychoanalytic Movement" that analysis presupposes "a situation in which there is a superior and a sub-ordinate."

Schiffer knew of this urge in me. He claimed it came from having a father who was weak in the very areas that formed the nucleus of my own pathology, that is, my promiscuity. If there was one "presenting symptom" as it is called in psychiatry, this was mine. In my eyes I had an insatiable love for women. To Schiffer, and here I think he had a point, I was just a womanizer. For Schiffer, as for Freud, hope lay in identifying with him, in taking him on as my own superego, my conscience.

Part of the problem, I knew already, had to do with a feeling I had that I could never fit in, was not "normal." Sometimes I yearned to just melt in, to fade into the crowd. I felt I looked peculiar, or talked differently, or did not dress right. When walking in a store or along the street or other "common" places, I would see what looked like an apparently "ordinary" woman, and I would think: no woman like that will ever want to be with me, not after I start to speak. And by and large I was right. Nevertheless, I had a compulsion (I see how easy it is to fall into medical jargon—all I really mean is an "urge") to prove myself wrong, and so I would often try to meet women who were obviously different from me. Dinners were often agonizing. I simply did not know what to say that would not give me away. I felt miserably self-conscious. When would they penetrate my disguise, know that I was a fraud? It was only when I succeeded in going to bed with such a woman that suddenly I would relax. The pretense was over, or rather it was no longer necessary. In sexuality I found my common humanity. I loved, I needed, to find that women were similar to me sexually, that they desired the same things, could be aroused in the same way. It proved my essential humanness, against all the odds. But it could never be satisfied. I never seemed to absorb the lesson, to feel vindicated. I had to do it over and over.

Whenever I passed a beautiful woman in a car, or on the street, I would experience the peculiar feeling of something receding, some possibility that was fading, something that was leaving my life and that I could not get back. I wanted to understand what this was all about. What I was hoping for, I suppose, was for somebody (Schiffer) to be able to say, "Oh yes, it is clear that when you look at this woman, this is what you are *really* feeling, or *really* thinking, only you don't let yourself know about it," and then to go on to give me an "interpretation" of what this was *really* all about. I regarded my "symptom" as not unlike a dream-text—it had to be deciphered. It contained some message from the unconscious, and only needed a person adept in its language who would be able to translate it for me. And then, when this had been accomplished, the symptom would miraculously disappear, and I would be cured. I would no longer have these peculiar feeling-states, and I would be free to move my life forward instead of forever worrying about what this regressive pull could possibly mean. However, I was doomed, it seemed, to stay on the surface. I talked of this constantly; Schiffer seemed to like this topic. He especially seemed to like one repetitive conversation we had. It went like this.

"I suppose my difficulty around fidelity has to do with wanting to imitate my father," I would say. "He was never faithful, and claims that his father hadn't been either. It's in our blood, he says. There is nothing in life as wonderful as a beautiful woman, and when you see such a woman, it is only natural to want to go to bed with her. If this is a disease, then I will be sick for the rest of my life, just as he is. There is nothing for it. Marital deception is the only decent thing to do. I must have lovers, as he had (he called them mistresses) and never let my wife know. But I want very much to get away from this model."

"I see your problem," Schiffer would reply. "Your father was no good."

Masson: "Don't you struggle with infidelity?"

Schiffer: "No. And you know why? Because my father was faithful, and so I have always been faithful. I had a good father. You didn't." It was one of Schiffer's maddening generalizations; the information I had given about my father, and my attachment to him, was very complex. Schiffer narrowed it down to whatever seemed to interest him at the moment, and this then became the whole picture. He would never acknowledge my father's many valuable qualities for which I loved him. He was not even interested in the nuances of my father's infidelities, some of which was mere talk.

"Now, let me construct a scene for you, to show you the difference between you and me," he would continue. "We are on a yacht together, in the Mediterranean, since you are such a fucking cosmopolitan. The French Riviera, where your parents hang out. On that yacht is the prettiest woman I know, Jackie Kennedy. She comes over to you, reaches out her hand for your prick, and you are lost. But then she comes to me. And what do I do? Not what you do, Masson, you can be sure of that. No, not me. Because I can remember my father. So I gently, very gently now, unwrap her fingers from around my penis"— note that I have a prick; he has a penis—"hold her hands, look her directly in the eye, and tell her, 'I love my wife and four children too much for this.' Then I walk away."

It was an impressive performance. He sincerely believed it. Or so it seemed. But how could I know what he would really do? I dredge up all of these shameful events from my past, and he meets me with a well-mannered cliché. True, I felt shamed. I thought, He's right. Why am I such a prick? Why couldn't I do what he does?

It seemed that at the bottom of Schiffer's theory of how therapy cures was the very simple idea of imitating your therapist. He often said it bluntly. "You have to learn to be more like me." But even if I knew who he really was, and I didn't, I couldn't. Since he seemed to believe that a good person can do no harm, anything he did to me was by definition good, since he considered himself a good person. Since I was not yet a good person, anything I did was by definition bad, and wrong. He had no need of ethics, he said, since he automatically did the right thing. He quoted a phrase that Freud had used in a letter to the American psychiatrist James Putnam, about ethics being self-evident. Freud claimed that he had never done a bad thing in his life, and Schiffer seemed to be claiming at least some of this same high ground for himself.

But the evidence of his great success in analysis was yet to become apparent to me. My good friend Mike was mysteriously silent when it came to praising Schiffer. It was as if he would be able to tell me what he thought only when I was through with my analysis. Indeed, this is exactly what happened—and in later years we were able to compare portraits.

In any event, I was not prepared to accept the testimony of somebody else. I wanted my own direct proof. After all, I was seeing the man five times a week. Could I not get a sense of who he was, what his character was like, from my own sessions? That was a difficult question to answer. Because I was so eager to believe I was being helped by a talented, ethical, benevolent, and intelligent man, I sought evidence for this wherever I could. Anything less than this was too dreadful to contemplate. Not only was my status as a candidate involved (as Schiffer made clear to me in the threat that he was quite prepared to suggest that I be "terminated" from the analytic program), but so was my own person. Nobody, he claimed, could handle me except him. It was my good fortune to have found precisely him, the only competent analyst in Toronto. Otherwise I could kiss my marriage goodbye. If I really wanted to remain faithful to my wife, he was my last chance, my only hope.

Then, too, there was the matter of the amount of money I had already spent on this analysis. Seventy-five dollars an hour, five days a week, comes to a lot of money. Admitting failure would have been costly on all fronts: for my self-esteem, for my professional career, for my personal life, and for my finances. It was easier to believe, even against the evidence of my senses and my intelligence, that the fault lay with me. None of us was truly free. All candidates are more or less in this position. When psychoanalysts boast that the goal of analysis is freedom, they cannot be referring to analytic candidates.

At a theoretical level, I wasn't sure I could believe that one could develop "character" by simply having a good model, by imitation, as it were. Even if Schiffer were to be perfect, it did not follow, for me, that I could be like him. How would it happen? By osmosis? Even in the case of fathers, the position that one imbibed character structure through observation, while widely believed and not implausible, was not without problems. There was room for doubt. And whenever there was room for doubt, I used it.

But the thesis became considerably less plausible in the case of an analysis. I was thirty years old. Surely my character was formed by now. I did not despair of making some changes, even essential changes, but I did not see how simply drinking in Schiffer's presence would substantially help. Even in the much more intimate case of a marriage, couples change slowly and gradually, even when convinced that it is necessary and vital to their marriage. Merely seeing Schiffer one hour a day, even for years, would not automatically make me into Schiffer. But of course if Schiffer were convinced that it was necessary that I become more like him, it was then self-evident that any amount of disparaging I did of him was ultimately bad for me, hence to be discouraged as self-endangering. I was not convinced, and did not want to be like Schiffer, because I did not know who he really was. I only had his word for it. His diagnosis for himself seemed to be a string of moral superlatives. I never did get a diagnosis from Schiffer that related to me, but I suppose he thought I was suffering from a narcissistic personality disorder. My effort to belittle him, then, was only a self-destructive urge. It had to be countered. That this was *self-serving* on his part I could easily see, but I thought at the time I had no alternative but to accept it.

Some of Schiffer's preoccupations had a universal ring to them, but they happened not to be mine. For example, he seemed to be much concerned with the question of trust, of who could be trusted. There were a number of times when he would slip this question into the middle of an ongoing discussion when it seemed completely foreign to whatever I was talking about. It would just lie there, inert. I always felt slightly embarrassed, and politely picked up the theme. But sooner or later I would run out of steam, that is, I could no longer associate it to my life. Schiffer was only too happy to pick up the slack and associate on his own: "I don't trust you or your wife. I trust Abrams,* but not you." Stunned silence. Again, maybe he had a point. Dan Abrams was, in certain areas, extremely trustworthy. But so what? What had that to do with me? Abrams was another patient; trust was not on my mind at the moment, nor was Abrams. Why raise it now? The problem, I thought, was Schiffer's, not mine. But now I had a thought and the basic rule of analysis was to put it out there: "Dr. Schiffer, I'm not sure how to say this—but while I think you have touched something worthwhile and even significant, perhaps this is your thought and your preoccupation, not mine. Anyway, you are here to analyze me. Whether you can trust me or not is irrelevant and shouldn't even come up."

"Don't tell me what I'm here to do. I know my job. And don't tell me what should or should not come up. If it comes up, it is to be discussed. Period."

"Yes, but it didn't just come up. You brought it up. You, not me."

"Because you're too chicken."

"How do you know?"

"Easy. Look at the way you're trying to worm out of this."

I couldn't win, so with a sigh I settled in and began talking about why I thought Schiffer thought he couldn't trust me. What I feared, though, was that all he meant by this was that I wouldn't be on "his" side when I left analysis and entered the world of Toronto psychoanalytic politics. I thought it, so I said it. He couldn't see why that wasn't a valid point. "Because, Dr.

Schiffer, frankly I couldn't care less about Toronto psychoanalytic politics."

I don't doubt that Schiffer was a man of strong feelings,
and that he was genuinely fond of me. But I believed that part
of his reason for liking me was the hope or, stronger, the expectation that I would be bound to him in the years to come.
As an analyst in training, I was a colleague, but I could also be
a potential ally. Schiffer and many analysts seem to see themselves at the center of a small circle of their ex-patients, all
analysts analyzed by their technique (in Schiffer's case, he called
it "the Schiffer method") who promoted their particular ideas
and ideals. Schiffer seemed to like the fact that I was in the
university. He sought, and eventually got, his own program in
psychoanalysis within the university system. Like many medical
people, he had an overblown conception of the importance of
the university and the trappings of scholarship. But it was no
use lying there speculating, because Schiffer always seemed to
have something of his own on his mind.

"Cannes."

"I beg your pardon?"

"Cannes. How do you want me to say it? With a French
accent? You know what I mean. You and your parents and their
South of France life. You think you're too big for this town,
don't you?" I had forgotten his code word for everything he
disliked about my family background. He seemed to hold it
against me that my father was born in France and had lived for
some time in Cannes, on the French Riviera. Once, in a cre-
scendo of anger, he told me that he had first landed on the
Italian Riviera during the war. It was a significant place for him.
But the contrast with the decadence of today was too much for
him. Not that he had ever visited Cannes. Yet again, I seemed
to be up against his ignorance, or prejudice, or just foolishness.

There was no way around it. The limits of this particular
human being were staring me in the face, yet to face them
squarely raised too many issues I preferred not to think about

concerning psychoanalytic practice. Did I not have my own limitations? Of course. Would there not be patients who would soon come up against them? Of course. Well then, did this mean I could not practice? My intelligence said "of course" but my prudence said I must have made a logical error somewhere. Moreover, I was fascinated with the transparency of Schiffer's vulnerability as a person. It gave me some strange comfort that the man analyzing me seemed to have so many problems of his own, and that they were so quickly visible. Sometimes I felt sorry for him, and protective—even, oddly enough, protective vis-à-vis myself. I did not want to hurt his feelings with a cutting remark that might hit home. At the same time this made me feel guarded, and guardedness is not something you are supposed to indulge in psychoanalysis. Things were becoming more and more confusing, from year to year. The more I knew of the actual practice of psychoanalysis, the less I felt I understood.

Chapter Three

GRAND ROUNDS

I had seen Schiffer for a year of personal analysis—really a preliminary observation period meant to function as a kind of window into the future course of the analysis—before being told of the decision upon which I felt my entire future depended: "You have been admitted into psychoanalytic training. Please report to the Clarke Institute of Psychiatry on September 15." Many of the analysts who were teaching the classes were connected, in one way or another, with the Clarke Institute, and so entirely for the sake of convenience we had our classes there. The fact that the Clarke was part of the University of Toronto Medical School no doubt exercised a certain pull on the analysts. The Toronto Psychoanalytic Institute itself had no formal connection with the university, a practice that goes back to the days of Freud, when no university wished to be associated with this fledgling and suspect science. Freud, as a gesture of defiance, then decreed that institutes should never be beholden to a university and this has remained the case, by and large, today. The Toronto Psychoanalytic Institute had no building, and was merely an idea in the minds of the various psychoanalysts who were part of it, but an idea they took very seriously.

And so, on September 15, 1972, on a cold gray evening, I presented myself to the Clarke. The building was forbidding. It was a psychiatric hospital; this was immediately apparent upon entering. I spotted another man carrying a letter of acceptance and introduced myself. He was David Iseman, an M.D. who had only recently completed his residency in psychiatry.

He was my age, about thirty. He seemed shy and, like me, in awe of this first evening, so we made our way together down the halls. I had rarely visited a psychiatric hospital before, and I was struck by the patients who shuffled past us in the hallways wearing slippers and institutional gowns, with their vacant stares, ticlike movements of the mouth, and slow deliberate gaits.

"What do they have?" I asked David. "What kind of illness is this that makes them look so vacant?"

"They don't have anything. Or rather they do. That's part of the problem. What you're seeing is not from the illness, it's from the cure. They look the way they do as a result of psychotropic drugs. Some of them probably have tardive dyskinesia, a real scourge, and one that is entirely iatrogenic."

I was glad that I sensed a note of disdain in his voice. I also liked the sound of "iatrogenic," which means "created by the doctor." I had never heard it before. I was puzzled and pleased by his apparent irreverence for psychiatry.

"They're not suffering from an emotional disease?"

"You mean an affective disorder? Oh, I guess it's the same. You know, when it's put like that the physician in me bristles. When I was doing my histology rotation, I never saw an emotional disease under a microscope." He was thinking out loud, and was thinking critically. I realized I could learn from him.

I also now knew that David was no dyed-in-the-wool psychiatrist. We both smiled, and I sensed we were going to be friends.

When we got to the room there were eight other candidates already waiting. Seven of them, including one woman, were medical doctors, psychiatrists. One was a clinical psychologist, and I was a professor of Sanskrit. We were all in our thirties, all somewhat apprehensive at beginning what would be at least eight years of training, and all inclined to form a unit.

At that time in Toronto there were about fifty practicing psychoanalysts, but only six were training analysts, senior analysts qualified to see candidates in analysis. All of us were seeing these six analysts, and since these men were not only going to analyze us but also teach us and supervise us, all of us wanted to determine who was seeing whom, and why.

David and I quickly exchanged this crucial information. I told him I was in analysis with Irvine Schiffer, and he gave a start. "So am I. That makes us brothers, of a sort." I grinned. When we turned to the others and began asking, we found that nobody else was forthcoming on this topic. It was considered, like everything else having to do with analysis, a private matter. I was at that moment struck by the contrast between free association as the method (and partially the goal) of psychoanalysis and the very unfree association of psychoanalytic candidates with one another. Within weeks I knew whom everybody was seeing, but I was not told directly. And all the others knew, too.

Our first lecturer turned out to be Peter Thompson, the analyst-interviewer who had so kindly interpreted my dream. He merely welcomed us, told us some of the mechanics of the seminars, gave a short speech about losing our identity as psychiatrists, which applied to most of my colleagues, to take on the new one of psychoanalyst, and went on to speak of some of the differences between the two professions. We would not be using physical methods, no electroshock, no drugs. We would not be concerned with committing patients to hospitals. Our gaze would focus not so much on the "here and now" as on the past. We would be less concerned with "fixing" and more with understanding. It was the complexities of the human mind and the human emotions that would now engage our interest, not the diagnostic categories of traditional psychiatry.

I liked what he was saying. Two of my colleagues objected, however, saying that they were merely adding to their identity as psychiatrists, not changing it. Psychoanalysis, they stated, was one more tool to use in psychotherapy, one small part of psychiatry, not its replacement. They saw no opposition between the two, and Peter Thompson (who was a psychiatrist himself, as were all the training analysts—in fact at that time, all the psychoanalysts in Toronto, with the exception of one, were medical doctors and psychiatrists) good-naturedly made the point that none of us really knew enough about psycho-analysis to have a very firm opinion and that the matter could be disputed later in our training. He merely wanted to point out that much of what we would be reading, in the years ahead of us, was more like literature than medicine.

I was relieved, not least because I had a certain anxiety with respect to my background. My first-hand knowledge of psychiatry was nil. All the other candidates had already completed their psychiatric residencies. They all had had hospital experience, working with so-called "psychotic" or "schizophrenic" patients. I did not. Would this hamper my training? Later, when I myself raised this issue with the training committee, which consisted of the training analysts, several members thought it would. It was suggested that I, alone among the candidates, attend psychiatric grand rounds at two of the private teaching hospitals in Toronto, as well as offer to do some volunteer observational work at a large state hospital, Queen Street Mental Health Center. The prospect excited me, though David cautioned me that I would probably not like what I saw.

As I write these words in 1990 I can once again, nineteen years later, feel the excitement I felt then. A whole new universe was opening up to me, and I anticipated the discovery of a world of emotions until then hidden from me. I felt like an explorer on the verge of reaching a new continent. I was open, curious, intellectually insatiable. I could not hear enough, read enough, learn enough about psychoanalysis. Yet at the same time, the concrete world that was right in front of me was in contrast to my theoretical expectations.

"When a child in the family manifests gross pathology . . ." these words startled me into consciousness. They were enunciated, for emphasis, very slowly, and in a booming voice. There could be no doubt about it, the department chairman was a fine orator. He had acted on the stage. His voice, his urbane wit, his friendliness, his poise, his great knowledge of literature were all admirable. He laughed a great deal. He liked to make jokes. You had to like him.

But you did not have to like what he said. And I did not. What was it to "manifest gross pathology"? In this case, an eight-year-old boy was the "identified" patient. This word "identified" was a popular and venerable psychiatric term. He had been "identified" as the patient by his mother and father, simply because he was not doing well at school, he had few friends, and he was a "problem" at home. How was this, I wondered at the time, "gross pathology"? Where was I? I was at grand rounds.

When you thought about it, which you weren't really supposed to do—the very fact that it existed was sufficient reason for you to accept it as ethically correct—there was something more than slightly odd about the very idea of psychiatric grand rounds. The entire hospital staff would meet once a week in an auditorium, and one of the senior psychiatrists would present a case, either literally by bringing the hospitalized in-patient along, or by presenting a case from his or her hospital practice. There seemed no way to preserve the dignity of the poor patient, who had little choice about whether to appear. Grand rounds was patterned after the other medical specialties, where the senior physician would visit patients on a given ward with his students in tow. It was meant to be instructional. But the human mind is not like the human liver, the kidneys, the pancreas, the spleen, or any of the other organs of the body, and this became acutely obvious to all of us on every single occasion. The mind is an abstraction, not an organ. And when we talked about emotions, which is what most of these sessions were about, then we were not talking just about the brain. Wasn't this true? Would we not smile at the French psychiatrist, J. E. D. Esquirol's phrase, "a lesion of the imagination?"

Most grand rounds were evenly divided between the

"organicists," who focused mostly on drugs, but also on e.c.t., electroconvulsive therapy, and even, ominously, on "surgery," and the psychotherapists. Although I was completely out of my depth when it came to discussions of medication, I listened with a healthy dose of skepticism. Because sometimes these same doctors also presented cases of therapy. It soon became apparent that every presentation of therapy was only as good as the intellect and heart of the presenter. You did not, you *could* not, learn about the patient, but you learned plenty about the presenter. Especially about his or her limitations (actually I shouldn't bother with the "his or her" formulation, since with only one or two exceptions the presenters were male). It was a bit like reading a bad novel. Some depth to the characters was missing, and the blame could only be laid to their creator.

So here was a department chairman talking about still another "patient," Jill, nineteen, "who was admitted to the hospital with a schizophrenic psychotic decompensation." But what was this, really? The chairman told the audience that the "mother was a cold angry woman who had never been fulfilled by her weak, infantile husband." The words resounded in my mind. These were not medical words, these were the words of judgment, the words of conversation, of novels (even if bad ones). What place did they have in a hospital, in a psychiatric hospital, no less, where the words that a psychiatrist used to describe a patient were taken absolutely at face value? They became, by virtue of his using them once, embedded in stone, the rigid irons that would bind this family. This was not medical science, this was the politics of psychopathology.

The chairman was known to the community at large as a person who defended the essentially medical nature of psychotherapy (it is, he had said in print, "an integral part of medicine"), as long "as the uncertainties of psychotherapy are submitted to research and evaluation." But while he was respectful of psychoanalysis, he was not himself a psychoanalyst. He was a psychiatrist. He was a scientist and had published many papers in which he measured things like how children responded to hearing their parents' voices. He certainly had nothing against physical therapies for what he called "psychological disorders." He had written that "somatic intervention [e.g., electroconvulsive therapy, or e.c.t.], may change a purely psychological disorder." But with us he did not ask any of the larger questions, and we were encouraged not to ask them either.

How did we know, for example, that somebody was "sick"? It was simple: they were brought to the hospital. The chairman made it clear that a person who had been "identified" as a patient by the family, was, in fact, disturbed in a psychiatric way. People apparently did not err when it came to making these kinds of home diagnoses. Thus he told us, speaking of the "maladjusted" (a medical term?) child, that we should accept

> that the "identified" patient is "sicker" than the others. A study by S. Wolff (in the *British Journal of Psychiatry*) lends support to the family's identification of its most disturbed member as the "sick one." She compared children brought to psychiatric clinics with controls in the population at large and found on the basis of tests and interviews that referred children were psychiatrically disturbed and that their referral is not arbitrary, nor simply a matter of the family identifying one of the children as sick without sufficient grounds.

To me, this was suspiciously convenient for the psychiatrist.

What gave the psychiatric community this power? I believe it begins with a fundamental claim: that of being able to distinguish between a "healthy" or "normal" mind and a "sick" or "abnormal" mind. In actual fact, not all that much had been written on this basic distinction—it was felt to be more a question of "feel." The few books and articles that did exist were hesitant, tentative. It was clear that the authors who did write about it, people like Georges Devereux, the ethnopsychoanalyst, Lawrence Kubie, the psychoanalyst, and Norman Reider, a prominent psychiatrist and analyst, felt they were on slippery ground. Without this basis to legitimate what we were doing, grand rounds—indeed, the entire psychiatric and psychotherapeutic enterprise—was suspect.

Maybe, then, the purpose of our weekly meeting was simply to offer one another assurance, maybe it was no more than a bonding ceremony, like primate grooming, telling one another that what we were doing was benevolent and necessary. But each and every time I attended grand rounds, my doubts grew deeper and deeper. A fantasy kept gnawing at me: one day, at an appointed hour, all the psychiatrists would suddenly look up as if awakening from a bad dream, would take off the white coats that many wore, perhaps to make them feel more like medical doctors, would unlock the wards, embrace their patients, and walk away from the hospital, never to return, muttering to themselves that the whole idea had been absurd to begin with. But the only person I was to see walk away was myself.

Grand rounds did nothing to alleviate my ignorance or sense of hopelessness. The psychiatrists seemed to be as shut away from these private worlds of pain as I was, only they were usually more reluctant to admit their ignorance. So the case histories tended to be short on personal detail and on experience, and long on procedures. The psychiatrists spoke of what they did *to* the patients as opposed to what they heard *from* them. The major treatment was mind-altering drugs. No doubt these drugs masked the symptoms of illness, but they also completely ravaged the personalities of the people who took them. There was no black market or street trade in any psychiatric drugs—every patient avoided them like the poison they were.

By and large these patients were not middle-class. *They* were different from *us*. The class system of course extended to the staff as well. Perhaps this is why some of the nurses were kinder to the patients than the psychiatrists: they felt closer to them. Some, however, perhaps in order to emphasize or create a distinction, treated them badly. The patients had nothing like a union or an ombudsman who could speak for them, and when they were mistreated they had absolutely no recourse. Lawyers interested in patients' rights in those days were a rarity.

I began a voracious reading program in psychiatry. I could not get enough of the *Journal of Nervous and Mental Disease*, *The American Journal of Psychiatry*, *The British Journal of Medical Psychology*, *Bulletin of the Menninger Clinic*, *Psychiatry*, and so on. But no matter how much I read and how much I talked to the psychiatrists, the contrast between the self-evident misery I saw right in front of me and the inadequacy of the methods of dealing with it was overwhelming. Still, my knowledge of psychiatry was meager, and perhaps I was unlucky to see it at its worst.

When I volunteered at Queen Street Mental Health Center, as the training committee had suggested, I was assigned a young patient from South America because I was the only one there who spoke Spanish, and he spoke no English. I talked with him frequently. One day he came to see me in a very agitated state. "*Ayuda-me, por favor* (Help me, please)." The male nurse, it had turned out, had ordered electroconvulsive therapy for him and he was in mortal terror of it, having had it once before. I told him to calm down, that nobody could give him shock treatment against his wishes, and I would immediately speak to the powers that were. He looked at me with what resembled a look of pity, and I left to find the appropriate doctor. He turned out to be a mild-mannered, kindly psychiatrist who was somewhat bitter at being forced to work at Queen Street rather than in one of the more prestigious teaching hospitals associated with the university. I told him my story and he looked at me the same way my South American friend had.

"Look, Jeff," he began patiently, "you are new to psychiatry, and new to Queen Street. You have a lot to learn. Who gets electroshock and who doesn't is not a medical decision, it is a political decision, and depends on who has the power. One thing you can be sure of, the power will never lie with the patient, nor with anybody who wants to help him. Even if I could do something about it in this case, which I probably couldn't, it will happen again tomorrow, or the next day, or next month. To him, or to somebody else like him. You have just seen something for the first time that is routine here. If you think your complaint is going to alter the system, you are naive. If you don't believe me, go and speak to the director of the hospital and see how far you get."

What he was describing sounded more like a torture chamber than a hospital. It occurred to me that my friend was being punished, not treated. Yet the psychiatrist had said all this in a matter-of-fact way, with such a tired voice. How could he work here and live with himself?

"I *will* speak to the director," I said. I could not believe that such cynicism could possibly reign throughout the hospital. I was right. The director was not cynical, he was absolutely persuaded, he told me, that nothing could be better for my friend's mental health than to have a dose of e.c.t., no matter who prescribed it. Better it should come from a physician, but the nurses have been here so long that they have learned the danger signs and know when a particular patient requires the soothing surges of 120 volts to the brain. He had nothing to be frightened of; why, it was totally harmless, and whatever pain was felt was forgotten afterward, because the shock to the brain was so great that many brain cells involved in memory were destroyed, and nobody remembered the actual pain. He began musing about this paradox of human memory, what is pain if we cannot remember it afterward—and I knew I had lost my case. My short tenure at Queen Street was coming to a close. I went back to the kindly psychiatrist and told him that if the patient were shocked, I was leaving that day. I am afraid this did not have an exactly galvanizing effect on the hospital. I resigned a few hours later and never returned.

I was convinced from the beginning that traditional psychiatry had little in common with psychoanalysis, and from then on I wished to dissociate myself from psychiatry entirely. I was not alone in this; the institute itself seemed to be divided along those lines. The older analysts were more loyal to Freud, hence to psychoanalysis, than to their own psychiatric training, which, for some of them at least, was merely a means of becoming an analyst. Many of the older analysts, in their sixties or seventies, traced their own lineage back to Freud. Of course in terms of prestige, the closer one could come to Freud, the better. So training analysts whose own analysts had direct contact with Freud were at the top of the totem pole.

For Freud, psychoanalysis was a revolutionary activity, and to this extent it could not subordinate itself to psychiatry or medicine. Indeed, Freud argued passionately against the American desire to subsume psychoanalysis within medicine. From the very beginning, the American Psychoanalytic Association, in direct contradiction to Freud's wishes, refused to train non-M.D.'s; this was true until quite recently. (The situation is changing this year as a result of a successful lawsuit by the American Psychological Association to ensure that more nonmedical candidates, primarily psychologists, are accepted for training by the orthodox and prestigious Freudian training institutes.)

This was not the case in Canada or in most of Europe, where probably half of the practicing analysts have no medical degree. But most of the younger analysts in Toronto had been trained in America, many of them at the Menninger Institute, and were inclined to see themselves as psychiatrists first. Most of them worked part time in general hospitals, and were not that far removed from their own medical school training, at least in spirit. Also, it was not easy to make a living practicing psychoanalysis, and so it was necessary to supplement one's income with other psychiatric procedures, such as consultations, short-term therapy, even, in a few cases, dispensing psychiatric drugs. In America most of the chairmen of departments of psychiatry tended to be analysts as well, and "building bridges" between psychiatry and psychoanalysis was a popular activity (every analytic conference had at least one paper so titled). While it was true that many of the great names of psychoanalysis were nonmedical analysts (Anna Freud, Hans Sachs, Erik Erikson, Ernst Kris, Robert Waelder, Marie Bonaparte, Siegfried Bernfeld, David Rappaport, etc.), it was also true that Freud himself tended to favor the medical doctors, or at least to feel more at home with them. Freud never seems to have accepted his own daughter, Anna, as a "real" analyst. In 1926 he wrote, "My daughter Anna has devoted herself to the pedagogic analysis of children and adolescents. I have never yet referred to her a case of severe neurotic illness in an adult." As if that were somehow admirable.

I had been warned that as a nonmedical analyst I might begin to feel like a second-class citizen. And many times it had been hinted or openly suggested that because I lacked a background of patient care, I might do better to engage myself in pure theory or pure research rather than work directly as a clinical psychoanalyst. Since I was to hear the same comments from well-meaning people such as Muriel Gardiner, Anna Freud, and Kurt Eissler later in my career, I knew that there was a kernel of truth to it. But I took the kernel to refer to my own abilities rather than to a position with respect to medicine. I personally did not feel committed to "helping." I did not have what Freud called the *"furor sanandi,"* the healing frenzy. Neither, by his own admission, did Freud. But then perhaps this is what made Freud a better theorist than a clinician. At least this is what the rumors were: his self-analysis was not successful because he did not have a good analyst.

Nevertheless, I was skeptical of the argument that the best analysts were the worst theorists. Usually this boiled down to a simple formula: if an analyst had published nothing and seemed, in conversation, to be unusually obtuse, this was only because he was such a superb clinician. I learned that when somebody described a respected analyst as a fine clinician this only meant that he had never published anything. Once I made the mistake of asking one of our teachers, "How do you know that Dr. So-and-so is such a fine clinician?" There was a pained silence. After all, none of us had ever observed anybody analyze another person except our own analysts. And that was the way many of us learned what *not* to do.

Chapter Four

THE WORM TURNS

There were high points (few) and low points (many) in my analysis, but there were also days, day after day, of the middle ground, when Schiffer was not doing anything particularly obnoxious and the "process" of analysis took its course. I am reminded, yet again, of a marriage. "What was your marriage like?" sympathetic friends ask afterward, and how can you answer? You tell the anecdotes, the worst times and the best times that stand out in your memory, but you are left feeling that the actual marriage itself, the day-to-day of it, has not been conveyed. True, there is no typical day in the life of a marriage, any more than there is in analysis. Nevertheless, a routine was established.

I would invariably begin a session by talking about any dream I remembered from the night before. Since I so often felt sad, I would talk about this as well, trying to pinpoint what exactly was making me sad. Since my sexuality was a central issue, I would tell, in detail, about every temptation that came my way. "Why should he have to listen to this?" I would often think. "His life is so different," I thought. But was it? We were in that strange space that a prisoner and a guard can enter, where he was as much my captive as I his. He had to be there, every day, just as I did. And if I chose to bring into the room the atmosphere of gloom that I was locked into, he had little choice but to let it seep out and wash over him. There was little he could do, in any analytic sense, to protect himself, since it was part of his "duty" to allow me to develop a theme. And he often did allow me to do this. I don't think he liked it very much, and because I sensed this, I did try to provide him with pleasure.

I was relieved when the sessions were lively, even boisterous. I also felt more as though I was getting my money's worth when he was active and participated, even when I didn't like what he said. The times when he said not a word always shocked me. Yet I heard from other candidates that such silent sessions were often the norm.

Many analysts sat silent for hours at a time, following the analytic adage "When in doubt, shut up." The most common complaint I heard from my colleagues was that their analysts did not speak enough. In fact it was not unheard of for some analysts to go a whole week or more without saying a single word except "Your time is up now." A few were grateful for this silence, arguing that it gave them freedom to come up with their own agendas. But I don't think it is possible not to wonder what the person behind you is thinking as you are talking, recounting painful memories or admitting to embarrassing fantasies. Is your analyst appalled? Disgusted? Approving? Would he rather be somewhere else? Is he bored to tears? Is he distracted? Is he counting the minutes until it is over? A woman could, with justification, wonder if the male analyst is sexually titillated, or envious, or just plain out of his depth. In a quiet room where you have been speaking nonstop for fifty minutes, to suddenly hear "The hour is up" provokes a feeling of something incomplete and unsatisfying. Then you meet his eyes. And you wonder, and wonder. It is a strange way for somebody to pass the day.

For me it was made worse by the fact that by the middle of my years of training analysis I was already seeing patients, and I *knew* the kinds of thoughts that an analyst could have. I would play little games with myself to make the minutes pass more quickly. Did Schiffer do that, I wondered? And if he did, would he ever tell me? "Masson, I just can't wait for this hour to be over so I can get you the fuck out of my life and get back to my family." That would be fair, but he never said it. And I know at the time that I did not want to hear anything like this. But could I ever say such a thing to any of my patients, no matter how true it was? I never did. Schiffer, I believe, thinks that he came close to saying everything that was on his mind. But that is a matter of interpretation.

The struggle on an everyday level as to what was acceptable behavior on the part of the analyst never really got settled. There was the matter of the telephone, for example. It had always been my understanding, although Schiffer had never addressed the issue, that these fifty minutes were mine and should not be interrupted, except in a dire emergency. The first time Schiffer's phone rang, he immediately picked it up and there was a brief conversation. I voiced my sentiments about interruptions. Schiffer agreed. The next day, the phone rang again, and again Schiffer picked it up. No doubt, another emergency. But it began to happen almost daily, sometimes two or three times in a session, and not infrequently even more. I started to have my doubts that these were all emergencies, since I heard discussions about engines and turbos. I would lie there in silence, steaming. I knew the consequence of expressing my anger out loud.

Schiffer never apologized afterward, and I found it hard to begin again. Soon it became one of my major preoccupations; as in the case of being late, I felt myself getting more and more angry, to the point where I knew I would explode. Schiffer seemed to be answering *all* the calls that came to him, and clearly some of them were routine business calls (the word "portfolios" came up over and over).

Finally I did explode, and so, predictably by now, did Schiffer. To me this was more like a marriage (and a troubled one at that) than it was like an analysis. I made the mistake of telling him that I did not interrupt my patients by answering calls on their time, why did he permit himself to interrupt his patient, me, by answering calls on my time? It was not very logical, but he got the point. Only he didn't like the point and he charged, emotionally speaking, like a bull. "Patients? Your patients? Don't make me laugh, it hurts. When are you going to write up your collected case histories?" Sarcasm was his favorite form of attack, but it lacked something in analytic finesse.

He always seemed to go for what was underlying the anger, never for what had provoked it—understandably, since answering phone calls in a session was not something one would presumably wish to defend. I wanted him to acknowledge that his behavior was getting in the way of my expensive analysis. Again, we had reached an impasse, and I simply had to give in unless I was prepared for more drastic action, which I did not seem capable of taking. When he said that we could spend weeks on this unprofitable topic or we could go on to discuss matters of genuine importance to my life, I did not agree with the crazy logic that excused him from any responsibility for his behavior on the grounds that I should be concerned with my own analysis, since I was paying for these sessions and he wasn't, but felt he was right at least in claiming that there was no more point in talking about it.

Not only did he compete with me in terms of time and free association but also his interest struck me as more palpable when our discussions ranged over transference issues, where he might be a player. For example, he seemed less interested in my mother than he was in my father, although my relationship with her was no less complex or significant. Part of the reason, I felt, for his interest in my father was that my father was an easy cross-over point for him. Comparisons came to him quickly and easily. He did not mind talking about women when they involved my pathology, but he had far less time for my positive feelings about women. Perhaps simply because women were less like him. I was beginning to suspect that he liked thinking about himself and talking about himself more than he liked talking or thinking about me. Understandably. Except that I was paying him for something other than a mutually satisfying conversation.

Not that he was uninterested in me. He was interested,

as long as he was involved. He did not limit his free association to himself but extended it to me, my life, my dreams, my relationship with past figures. He was never, in his own opinion, absent from a single dream of mine. He always found himself. If I dreamed about a decrepit old man, of course it was he, and I was demeaning him. If I dreamed about an athletic young man, of course it was he, and I was identifying with him. If I dreamed about my father or any of my colleagues, it was obviously he. But when I dreamed about a decrepit old woman, it was also about him—my attempt to sexualize our relationship. When she was a beautiful young woman, then that is who I wanted him to be.

I make it sound rather mechanical. Actually, to give Schiffer credit, it was not. He had a knack for dream interpretation. He loved it, and would sometimes, once in a while, come up with a striking association of his own that would hit pay dirt. To my shame, I cannot remember the details of any single instance of this, but I know it happened, and I know that telling him my dreams was the most conflict-free time we spent together. He had a fertile imagination and a very developed capacity to free-associate. These are valuable qualities, and he possessed them both to a greater degree than I did.

Again, in all fairness to him, I should state that sessions were never dull or boring. When he was in a good mood, which was often the case, he was extraordinarily entertaining. I rarely felt that he would rather be doing something else, which is a lot more than I can say for myself. He was a passionate man, and I got the feeling that he had a fierce loyalty to at least some members of his family. He was highly emotional and seemed capable of great love. He was quick, and clever, and funny. But in attempting to understand another person's life, I thought he was in over his head. Not surprising. So is every other analyst who has ever lived.

Where I thought Schiffer was deficient, and where I shone, was the ability to do the digging, the analytic detective work, searching for buried memories. It was not easy work, but I enjoyed it, and was thrilled when I could lay the groundwork for a patient recovering an early, lost memory. But my talent utterly failed me in my own case. I found that as hard as I tried, I was not coming up with anything very significant. Partly, I am convinced now, this is because Schiffer seemed to me so casual about it, and so uninterested. For me, dreams were the royal road not to the unconscious, but to my lost past. This was not Schiffer's position, and so we never got very far. I would like to have known a great deal more about my childhood than I knew, whereas Schiffer was apparently always looking for closure, a way to make the bridge to the transference, make a snap interpretation, and be done with it. I found this frustrating. But at least our conflicts here were kept in some kind of abeyance. He was not ugly about it, just, I thought, uninterested.

Where our pathologies happened to coincide, though, the result was great pleasure and absolutely no progress ("How have you changed?" was the standard question after an analysis. Now I wonder how legitimate it is to demand such change in a person's life.) I was a gossip. I still am. I got pleasure from it then, I get pleasure from it now. Still, I would rather not have paid $75 an hour to exchange gossip with my analyst. I provided Schiffer with a great deal of amusement when I recounted the foibles of my colleagues and mentors. He laughed and laughed, and clearly loved it. In fact he said so, often. Once he made me swell with pride: "Yours is the one hour in the day that brings me pleasure. I'm glad you are coming now in the evenings. It gives me something to look forward to. We sure do have some good laughs together." True, he had a wonderful sense of humor (much better than mine) and loved to hone it on other analysts as well. So when I would provide him some good gossip about another ridiculous remark by one of his colleagues, he would howl with laughter and then proceed to tell me his own favorite story.

Once he told me in some detail what he thought of a politically powerful psychiatrist we both knew. Schiffer hated him. So, I thought, did I. Schiffer told me, with some glee, that this man hunted, knowing that I would dislike him for this all the more. But as he went on and on, I became a bit uncomfortable. I felt I was hearing too much, even if some of it was exhilarating. I said this to Schiffer. "You are the only person who knows what I really feel about him," he told me. This was a genuine compliment, and it was difficult for me to turn away from it. But combined with the feeling of privilege and honor was a nagging sense that this was all wrong. These were things I should not be hearing. It was a little bit like hearing about my father's escapades. It fascinated me, but I knew that in some very profound sense, this was none of my business. Where I stopped finding Schiffer amusing at all, and when I knew that some ethical barrier had been crossed, was when it came to his comments about other patients in analysis with him.

The first time he made a remark about a patient, I was genuinely shocked. I was silent. He remarked on it. "You didn't expect me to do that, did you? Hey, I told you. This is my kingdom. I rule."

"But Dr. Schiffer—you are betraying confidences by talking about your other patients with me."

"No I'm not, I'm just telling you what I think about them."

"That's true. But you have just accused one of your patients of having been an anti-Semite." It seemed that this was one of his all-purpose insults.

"Well, so what? He was."

"Possibly, but when I complain that my friend Abrams, whom I like so much, is dull, it is not for you to say that you agree."

"Glad you brought that up, Masson. Truth is, Abrams is dullsville. Plain dullsville. It's got to be said, Masson. He is. I know. After all, I'm his analyst."

"That's just the point, you have privileged information. You may or may not be right, but it is not my business."

This was unbearable. If he talked about them, why wouldn't he talk about me? In such a closed community, the thought of having your analysis the subject of mirth was not a pleasant one. The most alarming part of this behavior, though, was that he seemed genuinely unconcerned. It just struck him as a matter of course, not even worth discussing.

"Why are you telling me this? What are you trying to do? I know he is dullsville, but he is also decent, a kind man."

"Yup. Kinder than you. I agree."

"But why make these comparisons at all? Why tell me what you think of your other patients?"

"Because it's the truth, and I know you have a great respect for the truth. If I told you anything less than the truth, you would stop respecting me, and we both know that would be your doom."

Again, I felt he was twisting something to protect himself, using the "truth" to mask his own character flaws.

Sometimes I thought Schiffer's behavior bordered on the comically absurd, though this man, so humorous when it came to the foibles of others, apparently failed to see the humor in his own "craziness." But perhaps I simply failed to appreciate his humor. Small details were irrelevant to him in the face of the big picture. I remember the time he was ranting about my lack of bravery. "What evidence have you for this?" I asked, a little bit puzzled by the vehemence of his argument. "I have plenty," he said. "You know what I did during the Second World War, don't you?" Actually I didn't, but I thought he may have told me and I might have forgotten and I didn't want to appear insensitive. "Yes."

"Well, where the fuck were you then?"

"Dr. Schiffer, I hate to blow holes in your theory, but I have to remind you that I wasn't born then."

"The hell you weren't. I'm talking about the day that America entered the war."

"If I'm not mistaken, I was too young to volunteer; I was less than a year old."

"You know, Masson, your excuses are beginning to wear on me."

Candidates in analysis are not allowed to attend the scientific meetings until well into their analysis, for obvious reasons—watching your analyst perform is an enormous complication. In my sessions Schiffer had shown himself to be a mighty critic of some of his colleagues and their defects. I couldn't wait for the day I would be allowed to see Schiffer take on his enemies in public.

Finally the day arrived. Dr. Garbin, who was one of the senior training analysts, was to speak about "sexual enthrallment." I didn't like him and as far as I could tell, not many people did. He was always extremely formal, immaculately dressed, with a most precise manner of speaking. He was intelligent and well read, and not a very warm person. I could not imagine him speaking of sexual love.

The case he presented was less than enthralling. I was fascinated by the fact that in less than one hour, a person's life was being summed up. It was a risky undertaking, under any circumstances. And when Dr. Garbin read us his summary, in the somber tone he gave it, it sounded more like a judgment, a final judgment, than an interpretation, and I could just imagine how stunned, or stupefied, or mortified the patient would be to hear it:

"The 'truth' which dominated this patient's life," he said, "was her discovery that she did not possess a penis and so had nothing to feel important about or to show off."

What a wild interpretation, I thought, and wished I could have said so. Of course, if pressed, Dr. Garbin could answer that this was really not an interpretation at all, but merely a restatement of the patient's own opinions; he merely acted as the screen upon which she could project her deepest and most feared wishes. He was no more than a gigantic computer into which she fed data in the form of years of talking at random during free association, which was then processed in the light of his knowledge of the truths of psychoanalysis. And of course, one of these truths, one of the bedrocks upon which psychoanalysis was founded, was that girls suffer from penis envy. But Garbin had decided to organize her lifetime experience around this central interpretation. Perhaps, just perhaps, the limit was in her. But it did not feel that way. Hearing him tell it and cite the literature to buttress his opinions in his dry, methodical manner, I couldn't help but see again the limitations of this particular man—his own obsessions, his own concerns, the limits of *his* understanding.

People did not generally object during these occasions. And if objections did come, they were often so euphemistic that nobody knew what, really, was being objected to. Somebody did, on this occasion, use the word "arbitrary," and Garbin did not take kindly to the comment. He reminded his listeners that he was basing his "insights" on a vast corpus of clinical writings in psychoanalysis, with particular emphasis on the papers of Anny Reich, Wilhelm Reich's first wife, on narcissistic object choice in women, on extreme submissiveness in women, and on early identifications as archaic elements in the superego. The analytic buzzwords were flying now in both directions. Clearly the paper was a success and was well received; there were no further objections. Here was a room of fifty analysts, and nobody stood up to protest the vaguely hostile mold into which one woman's life was being crammed willy-nilly.

Adversarial comments were not welcome. This was no courtroom, this was a scientific meeting. We were being instructed. Once in a seminar Garbin had reminded the students that he was not there to refight all the old psychoanalytic battles. He had better things to do with our time. Some of this was directed against me, for being too "humanistic," and some of it was directed against one of the psychiatrists whose orientation was too medical. I was struck at the time by the impartiality of it. For many analysts, Freud was right, it was the truth. All objections were merely due to ignorance, or a lack of insight, or finally to resistance. Legitimate disagreement was impossible. You did not argue about two plus two. If somebody didn't see it, you made inquiries into their pathology. But every clinical conference made me see that the stakes were very high indeed. *Somebody's life was under discussion.* It was not merely a matter of *preferring* one explanation to another. It was even more important than a court of law. And for this woman, who was being discussed now, her analyst was more than a judge, he was God almighty. Garbin's own candidates were exchanging admiring glances with one another and nodding blissfully in Garbin's direction.

I looked around the room, until my eyes met those of my analyst. Schiffer smiled and nodded, and then rose from his seat. He was alone, but he was going after his man. Abrams and I grinned at each other, put our hands over our eyes as if to shield them from the bloodshed that was to follow. We need not have taken precautions.

"Steve," Schiffer began, obviously softening him up by reminding him that they were old buddies, "as usual, I thoroughly enjoyed your brilliant paper. I have only a few minor remarks."

Unfortunately, he was right. They were minor. "Steve, I am reminded of Hans Loewald's article in the *International Journal of Psychoanalysis,* published in 1962, 'Internalization, separation, mourning, and the superego.' Can you say a word about that?" Schiffer was showing that he was a good player, and a team player. Garbin was happy, for in fact he had used this source himself and could drone on some more.

The next day on the couch I didn't know how to convey my disappointment. After all that smoke in my sessions, the dragon had no fire in him, and he was toothless. Schiffer did not seem unduly upset. "Garbin's paper was a disgrace," I said.

"Yup."

"I pity that woman," I continued, "I feel genuinely sorry for her. Her truth has been boxed in, sealed tight, unalterable forever."

"Yup."

"I would not want to be a child of Steve Garbin."

"God forbid," agreed Schiffer.

"But Dr. Schiffer, at the meeting you just complimented him. You did not voice any disagreement. And if you agree with me, then you didn't tell the truth."

"The truth," he said rather smugly, "is reserved for this room. In the world, I play the game, and so does everybody who does not want to lose." He said it pointedly. I had no sense of what was proper, where and when to reveal what I really thought. One day, I was going to get myself into trouble. He was right. It was a fair criticism of something he had seen in me that I had not seen in myself. But it did not still my deepening disappointment in him.

My friend Mike Allen, Schiffer's former patient, was now a full-fledged analyst and a professor with considerable political clout at the university. Schiffer told me that he was worried that Allen would betray him in some way. A number of sessions were taken up with Schiffer's lack of trust in Allen. He seemed incensed that Allen continued to be my best friend in spite of his warnings. He maintained that continuing the friendship was my way of obstructing the analysis: "You all want to murder me and take my place." It was Freud's primal horde theory, and while it sounded mechanical in Freud, for Schiffer it seemed visceral. Schiffer commented repeatedly from his analysis of Allen, but his greatest vitriol was reserved for what he seemed to regard as Allen's cardinal sin: Allen thought he could act independently of Schiffer. For Schiffer this only betrayed Allen's inadequacies. At a more sinister level, he took it as positive proof that Allen wanted to replace him by incorporating him, by ingesting him, by ultimately becoming him. In Schiffer's estimation, Allen wanted to be Schiffer. "But he can't as long as I'm alive, and so he wants me dead. But I'm not dead yet. If you guys think I'm buried already, think again. I have news for the lot of you, I'm still alive, and intend to be for a long time. So go find somebody else to murder." Schiffer lived in a wild, slightly paranoid universe, but it was not unexciting. It was as if he lived every day on the level of the primitive unconscious. *Outside* his office, he seemed perfectly sane. Inside he was a wild man.

His rule seemed to be that any behavior he engaged in was automatically good analysis, along the lines of *"L'analyse, c'est moi."* Similarly, any behavior he engaged in outside analysis was good, too, and was good analysis and good for analysis. During one session, when he bragged about the fact that this room was *his* room and he could do what he liked in his room, I made the mistake of responding by saying that I was looking forward to having my own room too, and one of the things I intended to do was to have my dog, Mischa, a black standard poodle of whom I was very fond, in the room with me. I was in a mellow mood, and I thought Schiffer was, too. But this remark, harmless as it seemed to me, galvanized him into a fury. "What? What did you just say? Maybe I didn't hear you right. You didn't just say, did you, that you were going to bring your DOG into the room with you, no, tell me I heard you wrong?" I was puzzled. "Well, yes, what's wrong with that?"

"You make me sick, you know that? A dog, a goddamned dog. I swear to god, Masson, you try that and I'll have you kicked out of the institute, I really will, and this time I mean it."

Alarmed at the ferocity of his response, I replied, somewhat weakly, "But Freud had his chow in the room with him."

"I don't give a fuck what Freud did."

I tried to pacify him with a little bit of humor. "Anyway, I'm not Freud, right?" but he was angry, and not about to be amused. I decided to let the matter drop, but I wondered what it was about such a seemingly innocent remark that could have provoked his ire to that extent. I even found it intriguing. But it was also, as I knew even then, alarming. Whom was I in analysis with?

Extra-analytic contacts were frowned upon by all classical analysts. And Schiffer called himself a classical analyst. He disdained other analysts who met patients outside of analysis. One day, as I was reflecting out loud on this fact, I mentioned that my parents were in town. "They are?" he said, sounding very interested. "Yes," I said, "I had the fantasy that we would all have lunch."

"Let's all have lunch," said Schiffer.

"But Dr. Schiffer, isn't that against your rules? Isn't that a parameter, and one you don't engage in?"

"Naah, and anyway, what's a parameter? It's just a word used by Eissler. I'm curious to meet your parents. Besides, I like to keep you guessing."

We did have lunch, and I can remember feeling that this was vaguely illicit. Schiffer was charming, and both my parents liked him. Schiffer talked about my talents, and how he expected me to make a significant contribution to psychoanalysis. Naturally, this made me acutely uncomfortable, a bit like a little boy at a parent-teacher conference. It also, somehow, made the analysis the next day feel unreal. As if we had already finished and the rest was just a formality, though this was toward the beginning of my analysis, not toward the end. I felt a little ashamed to tell him some of the things I did, given his high hopes for me. What if I disappointed him? What if I did not live up to his expectations? What if my very next sentence caused him to change his mind? This was quite a burden to have to bear when you were talking to somebody five days a week, eleven months of the year, for five years.

Meanwhile, the problem I had come into analysis for was not getting any better. There was the time I struck up a conversation with an attractive woman at a bus stop, unmistakably an attempt to pick her up. I offered to meet her for lunch the next day at the university. She came, and when I told her I was in analytic training, she proudly announced that she was also in analysis, with one of my classmates. I was crushed and worried. Any affair I had anticipated was now out of the question. But it also meant that I had just committed a public act: my attempt to pick her up would now be known to my classmate, his analyst, and his supervisor, people who I knew held no great brief for me. I was in something of a panic the next day. Schiffer found the whole thing delightful. "You are in big trouble, Masson," he announced with glee. "You sure are providing me with entertainment."

Amusing as it was to him, and quite apart from the imminent danger of my being asked to resign, I was also concerned, of course, with the deeper issue of my promiscuous urges. Not only was this bad for Terri and for our marriage, it was also bad for me. Why was I always on the verge of infidelity? Why was it so hard for me to remain monogamous? While Schiffer sometimes attributed my problem to the bad example set by my father, he had an even simpler explanation this time. "Your wife is not beautiful. In fact, she looks like a horse's ass."

This comment would have been more shocking to me had I been completely unprepared for it. But I was not. For years now Schiffer had repeated, over and over to the point of total numbness, a joke in very poor taste whenever Terri's name and the question of my marriage came up. He invariably said, "Take my wife . . ."—long pause—"please." It was, he explained to me the first time, a Henny Youngman joke. The first time I dutifully attempted to free-associate to his joke, taking it as seriously as I could, though it repulsed me. I asked myself whether there was any truth to this: was I asking to be rid of my wife, was I trying to end my marriage, or at least sabotage it? The idea was not totally unworthy of consideration, even if the manner in which it had been expressed was. But Schiffer did not stop there. He used this same joke, over and over and over. After about the tenth time, I was disgusted and said so. But this did little to stop him from repeating it, like an automatic response whenever Terri's name came up. On this particular occasion, however, he definitely outdid himself. To be sure, by now I was used to his vulgarity and his opinions. But even by Schiffer's standards, this one was astonishingly crude. It was untrue, and it was an absurd metaphor that did not carry a meaning for me beyond being extremely distasteful at all levels. As soon as he had said it he started to add his usual. "You know, do you not, the old joke, 'Take—' "

"I know, I know, you have said it a thousand times. Please, not again. Suppose, that by your standards, Terri is not pretty, is that really so important?"

"Yup, all-important. I oughtta know. I married my wife because she was beautiful, and I have never regretted it. And you know what? She's not even Jewish. For three years after I was married my father refused to speak to me."

As usual, this was a great deal to take in. First there was the insult to my wife. Then there was the fact that this insult was coming from my analyst. Then there was the information about his wife, which I was not supposed to have in any event. And why had he suddenly told me she was not Jewish? Was this a confession or a boast? Even more troubling was his comment about his father. Surely I was not supposed to know this. But it intrigued me, and I even felt bad for him. I did not like the sound of this father, yet Schiffer seemed positively to adore him. This was the man he had constantly compared to my own father, always to the disadvantage of mine. But my father would never have responded to a choice of mine in this punitive fashion. I was, in a sense, delighted to know something real about his life, but I was also perplexed as to how I was to use this information.

I replied, "But Dr. Schiffer, those opinions you expressed are not mine, they're yours."

"No, I'm just putting your thoughts into words, because you're too goddamned scared to do it yourself. You're the one who doesn't think your wife is pretty, and it bothers the hell out of you." In other words, he was claiming that he was simply giving voice to my secret thoughts, or, in analytic jargon, he was giving voice to my unconscious. Presumably I *felt* these things, but couldn't acknowledge them. His giving them words was supposed to jolt them into awareness, to make the unconscious conscious. That was, wasn't it, the goal of psychoanalysis?

But even if Schiffer was right that these words of his represented my unconscious thoughts—it would be very difficult ever to verify—how could I fail to notice, from the tone and everything else, that these were things *he* believed? He was expressing his own views, his own biases, and I said so.

"No, you are not right. These are not biases, because these are truths. *Every* healthy man wants a beautiful woman for a wife." Schiffer's blatant sexism sometimes took my breath away. "As for intelligence, that is not so important. It's only because you're sick"—his favorite epithet for me—"that you admire intelligence above beauty. And it is tearing you apart, the unconscious conflict, that is. Because theoretically you admire Terri's intelligence, her brilliance, but at a deeper level you resent her for not being beautiful. That is why you are always tempted to be unfaithful.

"By the way, your wife's intelligence is not natural. In fact, I find it disgusting. Because I know what it is really all about. And so does every other normal woman. Normal women don't want to be with your wife. They can't stand her. And you know why? Because they can tell that she is using her brain like a penis. Her mind is so developed because she is so filled with penis envy. She is so desperate for a penis that she has created one in her head. Her brain. Her huge brain is nothing but a substitute for her desire for a huge penis. Your wife has a cock for a brain, Masson, and you're getting fucked." He chortled in delight.

I suspected from the way he said this that he believed it. It was a combination of the worst of analytic theory (penis envy) and the worst of his own personal prejudices against women. He said it with such passionate self-righteousness, that I knew

I was helpless against him. It could never become the subject of a rational discussion.

I often felt that Schiffer may have had a point about me, but to know for certain it would be necessary to disentangle what he said from his own biases, and this was impossible to do. He had no interest in examining himself, claiming that he had already done so long ago, and to his complete satisfaction. His analysis was entirely successful, he told me often, so his interpretations came from a deeper level than did other people's; and in any case he had never really needed analysis, given his father. Here he launched into a somewhat incoherent narrative of Felix Deutsch, his analyst, and Deutsch's "bitch" of a wife, Helena, who never could stand him.

I couldn't tell if he really believed this, or if it was just a strategy, but it depressed me. I was in my fourth year of analysis, and was beginning to believe that I would never get what I had set out to achieve: self-enlightenment, insight, self-knowledge, freedom from my promiscuous tendencies. However, there were things I *could* get from him, and I felt it made sense for me to modify the practice of free association, to stop berating him and complaining, and simply take whatever it was that he was capable of offering. To me he was crude, and he was biased, and he was narcissistic, but he was not a stupid man, he was not without strong passions and a certain shrewd intelligence when it came to the unconscious, and he had some insight into what was wrong with me even if that insight might be just plain common sense. God knows that I was no paragon of virtue. I was lying to Terri, even though I genuinely loved her. I was not ready to renounce sensual gratifications for domestic satisfactions, and I could not face this honestly. I did not want to shoulder boring responsibilities, and give up making the kind of intellectual discoveries that I felt made life worth living. I was an adventurer, and not just in the best sense, either. I longed for adventure, sexual, intellectual, or even emotional. But once I found it, I was already looking for the next one. I was definitely not good for the long haul. Something had gone wrong, or had never gone right, and I needed help in finding out what I could now do to remedy the situation.

The problem was that as the analysis continued and began to reach its conclusion, it seemed to me that Schiffer was less and less interested in analyzing me, and more and more interested in securing my friendship. Although flattered by this, I did not honestly know whether I wanted to be his friend. For being his friend meant taking his side in analytic battles that began to seem to me more and more pointless and trivial— for example, his version of Freud over Garbin's. I did not really care enough about these small-town fights. That they mattered so much to him diminished him in my eyes.

At the same time, I was still very much under the influence of my own idealizing transference to Schiffer. I am puzzled to this day as to how I could sustain it. No matter how crazy his opinions seemed to me, I maintained till almost the very end that he was a kind of genius. I felt we were direct opposites in many ways. For example, he apparently did not like feminists. Once he complained that women behaved as if "Freud had hit them in the crotch with an ax." What women really wanted, he said, was really quite simple: a penis, either permanently or on loan. I could see how silly and dated these views were when he stated them as general principles, which he often did, speaking of "menopausal nymphomaniacs" for example. But I grew weaker when he applied them directly to me. As soon as he made an "interpretation" I felt a kind of obligation to accept it as valid, or at least make an attempt to accept it, to try it on. But sometimes, even when his interpretations seemed to me to be just plain silly, I was unable to see through them, and would often marvel at his wisdom. Part of this was undoubtedly transference, but there was also a part that was simply fear. It was easier to accept than disagree and risk his anger. He had many "stock" interpretations. Whenever a patient asked "How are you?" he invariably translated this to mean "I trust you're sick."

He was fond of claiming, with what he took to be uproarious humor, that in many marriages farting was the only mode of communication, a sentence he used in his book.

In fact, while I was still in analysis with him Schiffer wrote two books: a psychoanalytic study of time and a psychoanalytic study of charisma. When I read them, I simply could not see them in any objective light. My normal standards had vanished. I read them and enthused over them for weeks. Terri looked at them, too, and was puzzled at my reaction; she thought they were poorly written, badly argued, and more like "vanity" books than genuine works of scholarship. Naturally, the next day I had to report this. Schiffer went into a towering rage against Terri. He seemed satisfied, though, that my enthusiasm was genuine and undying. That was all that mattered: "I care about your opinion more than anybody else's." This of course made it absolutely impossible for me to have my own opinions, knowing that it mattered so much to him. Now when I look at these books I shake my head in amazement. Such is the strength of the transference.

But what is the nature of this powerful animal called the transference? Is it entirely what psychoanalytic theory claims it to be, a relic from the ancient past, a transferring of thoughts, wishes, feelings from early figures onto present ones? I am not convinced. Because there seems to me to be a simpler explanation for my idealization of Schiffer: I desperately *needed* for him to be the man he claimed he was, the man I wanted him to be. My own sense of worth was somehow tied in with him now. If he was a fool, then I was a fool to stay with him so long. At a deeper level, the qualities I ascribed to him, which were often qualities he claimed for himself, were necessary for my survival. I had to believe that Schiffer was wise and kind and good and selfless, because only if he was would I benefit from the analysis. If Schiffer were not a good man, what could I take from him? Of what benefit would be our sessions? It was in my own interest to maintain the delusion. It was self-preservation, I think, as much as a projection of the past. It was also undeniably true that I did not understand myself. The less I understood, the more of an "expert" Schiffer became. No doubt he liked this idea. He seemed not nearly as keen on my coming out of this "infantile" stage and growing up. My theory was that the notion of my disillusionment with him was not pleasing to him, even though it was, in principle at least, part of Freudian doctrine. I certainly needed to be disillusioned about Schiffer, but only in the presence of a Schiffer who could permit such disillusion, just as the child can only become a person in his or her own right when the parents allow this separation. Young, we are vulnerable, and when a parent cannot let go, something breaks.

I suppose it was having my own child, Simone, that allowed me finally to free myself from Schiffer. The disillusionment occurred the day he spun an imaginative and elegant explanation for why I was crying in a session. My newborn daughter was lying in the hospital with acute jaundice, and the pediatrician had suggested a blood exchange. Simone's life was in danger, and I did not know if I could trust the doctors. I was terrified that she would not survive the procedure.

Schiffer, meanwhile, seemed blinded by a rule of thumb in psychoanalytic technique that *all* events in the analytic sessions are transference events. Hence, in his interpretation my tears were crocodile tears (his phrase) whose purpose was to convince him that I was a man of compassion and depth of feeling. (Incidentally, this interpretation could well have come from an article by James Strachey. It is a danger to be too well read in psychoanalysis; you will recognize not just the direction of the interpretations your analyst makes, but in many cases even the exact wording. Schiffer might have spared himself some time and simply told me, "Strachey, *International Journal of Psychoanalysis*, 1923.") The interpretation, however, was to me a clear case of professional narcissism. In truth, I was so caught up in my daughter's suffering that I probably wouldn't have noticed had he keeled over dead in front of me in the session. Schiffer, perhaps convinced that only analytic insight could help, insisted on making one interpretation after another, all of which not only fell far of the mark, I am convinced, but were hurtful as well. When I told him that I was hesitant to authorize the pediatrician to perform a blood exchange, he was furious, and claimed I was trying to compete with a "real" doctor. "He is a physician, and he knows what he is doing," insisted Schiffer.

"Do you know him?" I asked.

"No, but he is a medical doctor. He is only doing the best he can for your daughter. Trust him and stop your horrendous acting."

As things turned out, after I had a discussion with another pediatrician at the hospital, we refused the procedure and Simone was fine. But I lost a great deal of confidence in Schiffer's ordinary, human judgment. I didn't really care, at that moment in my life, what kind of analyst he was, I cared what kind of human being he was. I had to tell Schiffer that for all his preoccupation with trusting other people, I now felt that I could not trust him to behave with simple human decency in a moment of need. He was enraged. He always became enraged at any criticism, but his anger flared up most easily when I was no longer angry, just telling him what I really thought. "Your performance at your wife's labor was a disgusting farce." I suppose he had not liked the fact that I was present, and that we were practicing breathing exercises beforehand. I did not, however, rise to the occasion. I was weary of his complaints. "Your wife's breast-feeding is a farce too." I was silent. "She's not a woman." Silence. "She never got an analysis." Silence. "You're both sick."

I was not angry now, but I was curious. "And you? Were you never 'sick'? Weren't you in analysis too?"

"No, Masson, I was never as sick as you. I went into analysis because of questions having to do with the war. Could I send men to die? It shook me up. But I never had your horrendous pathology. You're not even loyal." There it was again. He was referring to his book. A patient of his, a professor of

English at another university, had written a favorable but not very intelligent review. When I remarked on this he flew into a rage. "I don't want intelligence, but loyalty. You can stick your intelligence up your ass," he shouted at me. Anyone who did not like his book was envious, he said, still shaking with anger. He was not at all convinced, he said over and over, that all his analysands, and in particular, Mike Allen and me, had his good at heart. But why should we have his good at heart? And why should I be loyal to him? What was our proper relationship? Theoretically, at least, we did not have a personal relationship. I was paying him money, to do what he did, we were not friends. What were the obligations of the analytic son to the analytic father? Whatever analysts say in theory, in reality, none of them, it seems, ever gets over the shock of seeing ingratitude on the part of patients.

The analysis ended in 1976, before I graduated. The "rule of thumb" was that a personal analysis, for the institute, lasted from four to five years. Many went on much longer, but I was, by now, eager to end. I did not actually feel "healthier" than when I had come in. I still had, basically, the same problems. But I was much wiser now about how analysis worked. If I wanted to leave, I had to convince Schiffer that I was now "cured" of whatever illness he thought I had. The tone of the analysis had now to shift into the elegiac mode of saying good-bye. In analytic jargon, I had to deal with the trauma of termination. I cannot pretend, now, that there was not an element of sadness in leaving Schiffer. I had seen him every day for five years. I had gotten to know him, even against my will, pretty well. Mixed in with my distaste was a kind of archaic affection. For better or for worse, we had been close. He was like a parent—no matter how different you become, you can't help but feel a certain pang on leaving home. Becoming independent also meant for me that I could not really hate him, or want to take any kind of vengeance.

As for Schiffer, he had told me that in his eyes, I had "graduated" from his couch. He thought I had had a good analysis, and he was prepared to say so to the analytic committee that would be in charge of my graduation, which would occur after I had completed all my supervisions. I certainly believed him.

In fact, the analysis was only finished long after it had ended formally. I was wary of Schiffer now, but I had lunch with him from time to time. Reality dawned on me at one of these lunches. I was moving to Berkeley and had applied for membership in the San Francisco Psychoanalytic Society. They had invited me to give a paper. I discussed it with Terri and we decided to write something together about the sexual seduction of children. We tentatively called the paper we intended to write "The Navel of Neurosis: Trauma, Memory and Denial." I told this to Schiffer over lunch. I felt convinced of the truth of the ideas I wanted to put forth in this paper, even though I knew they conflicted with Schiffer's ideas, and the ideas of many analysts.

Schiffer asked me to tell him some of the thoughts Terri and I had talked about wanting to include in our paper. So I told him. I said that we believed (wrongly, as it later turned out) that Freud would not have been in sympathy with a relativistic view of what constitutes the truth of the past. We disagreed with the standard analytic position that there were only *ways* of apprehending reality. We thought we could find an underlying basis to the repetition compulsion in the strange bond that exists between a person and his or her past suffering. Traumatic experiences are always at least in part repressed. The repression also contains an upward thrust, so that one can even speak of a "buoyancy" of the repressed. The need to remember, then, is also a self-curative gesture, and people who suffer from lacunae of memory are preoccupied with attempts to piece together their own past. Linking this to psychoanalytic theory, we thought that the heart of the transference neurosis is its undoing, its unwinding into the past, carrying along, in its backwash, the compulsion to repeat. These repetitions were attempts to seek help and rescue, and therefore, interpreting

material from a severely traumatized person as fantasies ob-
scures past realities and is felt by the patient as a misunder-
standing of the most important events of the past. These
interpretations, which can be felt as prohibitions against memory, involve a form of deceit against which the patient is as
helpless in the analytic situation as he or she was in childhood.
The assumption on the part of the analyst that nothing out of
the ordinary has externally happened to a person living out a
compulsion to repeat in the transference neurosis, or that what
happened internally is, in terms of pathogenicity, far more important, would appear to be a derivative of the defensive illusion
that the world is benign.

I took a breath. I had finished. There it was, in the technical
terms that analysts use with one another. But it was, I thought,
unmistakable. I was telling Schiffer, "Look, here is what your
analysis has taught me. I have used it to know what not to do.
Every fault of yours has become a theoretical building block for
me. I know what to avoid. I have learned about psychoanalysis
by the caricature of it that you provided me."

Schiffer had silently been eating his soup while I talked.
When I finished he wiped his mouth and said quietly, even
with a certain dignity, "You're a thief, Masson. That's my
paper." I laughed appreciatively. Such irony. But he wasn't
smiling.

"I'm serious. I wrote that paper. You're stealing it from
me."

"Wait a minute, you must have misunderstood me, Dr.
Schiffer. I haven't even written the paper yet. It's about Terri's
experience in the Warsaw Ghetto and my ideas about sexual
assaults in childhood, and Freud's in my view mistaken abandonment of this theory. It has nothing to do with your interests,
which are, I know, quite different. I don't even think you would
like the paper; you certainly could not agree with the ideas in
it that I have just told you about."

"I told you, I wrote that paper with you. Now, you can
either acknowledge me as the coauthor, or . . ."

I was waiting. Or what? What could he possibly have in mind? I was about to be graduated, and accepted as a member of both the Canadian and the international societies, and I was getting ready to leave the city. What could he mean?

". . . or I will tell the membership committee that you have not finished your analysis, that you have to come back in, and find out why you want to rob me. You have a choice: put my name on that paper or come back into analysis."

The quiet manner in which he said this was in direct contrast to the wild content of what he was saying. I couldn't quite believe I was hearing him right. And yet, to my astonishment, I was not downcast at all, as one might have expected. I was exhilarated. What was happening? Suddenly it dawned on me: I had been right all along in my analysis. I had not imagined it, I had not fantasized it. Schiffer really did seem to live in a strange world of his own making, and he apparently wanted me to join him at any price. This was one club I had no intention of joining. I was wrong to think Schiffer had allowed my analysis to end normally, knowing all along that we stood on different sides of the canyon and could only shout to one another. He had done, I thought, the dignified thing, and let go. I was wrong. He seemed to have no intention of letting go so easily.

I wish I could say that I had the dignity to tell him to go to hell. Instead I told him I needed some time to consider his "proposal." I went home and called Mike Allen who told me I must have imagined it. But he had lunch with Schiffer himself the next day, and confirmed what I had told him. "I can't believe it. You're right. I guess Schiffer has finally gone off the deep end," he exclaimed. He advised me to call him and accept his offer, then forget all about it and do nothing. I imagine he knew Schiffer well enough to suspect that he wanted only the acknowledgment, not the reality. In truth, I don't believe he ever intended to actually blackmail me for his own gain. Rather, for some unfathomable reason, he seemed to want his name alongside mine on a paper. I did what Mike suggested. Schiffer seemed mollified, and indeed, I never heard about the matter again, though of course his name never appeared on the paper.

In retrospect, I am not convinced that my analysis was different from that of other candidates. Was Schiffer really so much worse than others? In some superficial ways I believe he was. For example, most analysts do not bully their patients about time or interrupt sessions by answering their phones. But in at least one important way he probably was not so different. All of the analysts had their blind spots, which they acknowledged by giving them a technical name, "the countertransference." But over the years I had met all of the training analysts, and watched them act in the real world. Each had what I considered a special prejudice, a weakness, an area of ignorance, a character flaw. And yet all of them thought it legitimate to offer themselves up as models upon which their individual candidates should pattern their lives. Some of them were anti-Semitic, some were racists, some were sexists, some lacked a generosity of spirit, others were dull-witted, some were mean-spirited and a few were just awful human beings. Yet each was a training analyst. Each had the right to train other people to become psychoanalysts, and could pronounce them either fit for this activity, or deficient. I did not put myself above these people, and I did not think that, on balance, Schiffer was either better or worse. But did any of us, any single one of us, have the right to judge another human being, believing that we had access to some great wisdom they lacked, merely because we had been analyzed? It was beginning to seem preposterous.

Unfortunately, as I was constantly being told, the core of any analytic training program was the quality of the personal analysis. If that went well, if it was a "transformative" experience, then, just like being in love, nothing else mattered. Yet I felt my analysis had been, to put it bluntly, a failure.

Had I not been a candidate, had this been a regular therapeutic analysis and not a training analysis, would things have turned out otherwise? I think so. I don't believe that Schiffer, or any other analyst, could have allowed himself the kinds of liberties with a "normal" patient that he could with a candidate.

We were a captive audience. Every one of us. I did not hear of a single candidate changing analysts, or quitting. A regular patient, if he doesn't like what the analyst does, can quit, or threaten to quit and mean it. The analyst can lose a fifth or sixth of his income in this way. The ties that bind can no doubt be strong, but at least they are not accompanied by the same devastating power to make or break a career that every training analyst has over every candidate.

The shadow that such power casts could not simply be burned away. It affected every aspect of the analysis. Being in such an analysis is like growing up with a despotic parent. There is no such thing as a benign dictator, because the power base for such a dictatorship is created at the expense of the subject. To be meek, abject, and obedient is to be robbed of some precious right to equality. Nobody will give this to you. You must take it. The possibility of analysis in an institute is an illusion. The only thing you can do with an illusion is to shatter it.

Chapter Five

THE MEN'S CLUB

1973

It was considered one of the major steps in a candidate's progress in the psychoanalytic program when you were given written permission to see your first case in analysis. It was then your responsibility to find a suitable patient. For the male candidates this meant a female "hysteric"; the one woman in our class had to see a male hysteric. This was the official rule; the reason behind it was never made clear, but it probably was because this is how Freud began. You were to see the patient five days a week, on the couch, and you were to be supervised once a week by one of the training analysts, for which you paid the equivalent of an analytic session, generally seventy-five to one hundred dollars. In order to graduate, you had to have a minimum of six hundred hours of supervision. But the hours were not automatically tallied. Your supervisor had to agree to count them. If the experience with your patient was not sufficiently exemplary, you had to go back and begin again. I know. It happened to me.

My first patient was a business professional. She paid me one dollar per hour. The low fee provided the candidates with a ready pool of patients; it also possibly represented what the analysis was worth—certainly this was a common perception on the part of many first patients. I was terrified of beginning an analysis. After all, I was still in the initial stages of my own. It had actually been going on for two years, but I still felt very much at sea. I was not making stunning progress in unearthing the sources of my own unhappiness; how could I help somebody else?

The analysis did not get off to a promising start.

"Dr. Masson—though I'm not even sure you are a doctor—how come you have such a small office, and what is it doing in the department of Sanskrit?"

It all sounded ominous, and I really didn't know whether I should make an immediate interpretation (e.g., "Perhaps you're afraid that I won't really be able to help you"—when this would have been a perfectly legitimate fear), or a "deeper" interpretation (something fatuous like "You are afraid that my penis is so small it is not worth having"—an interpretation style much favored by the many Kleinian analysts in the Toronto Institute), or just tell her my whole story straight out.

Before I could say anything, however, she went on to say, "I must warn you. I have been in analysis twice before, once in Europe for five years, and again in the United States for two. Neither helped me." She should also have warned me that she was saying more now than she would ever say at one time again. For it turned out that her problem, or rather mine, was that she lay on the couch for the entire fifty minutes in almost complete silence, day after day, week after week, year after year. I attempted interpretations, cajoling, begging, insulting, all to no avail. The breakthrough came, I thought, on the day she lay on the couch and proceeded to tell me that she had dreamed the night before. This was several months into the analysis, and she had never before volunteered anything, let alone something as tantalizing as a dream. My heart leapt, but I said nothing, afraid of frightening her back into silence.

"I was in a room," she began. Or so I thought. Alas, I was wrong. It was the end. That is all she could be persuaded to recount of the dream, or all she remembered, or all she dreamed. She had been in a room. No doubt the dream was not without meaning. It immediately occurred to me that this one sentence encapsulated her problem with me and the analysis: she was in the room, and that was all there was. She was there in body, but where her spirit was in hiding, I had, and still have, no idea. I told her all this. She remained silent. Not even a sigh.

No matter how unpleasant the sessions themselves were, they were far preferable to most of the hours of supervision. There is something particularly unsatisfactory about telling a senior analyst what you did, said, and felt with your own patients. The raised eyebrows ("You did?"), the smile ("Poor fool"), the grunts ("Never mind what I would have said, let's just get this over with"), all indicated your lowly status and pointed up your perpetual ignorance.

My supervisor on this difficult case, to whom I paid a hundred dollars every time I saw him, which was once a week, was Michael Goldfarb.* He had very little to offer me in the way of advice or interpretations, but he was quite methodical and never missed a single supervisory session. I read everything ever written on the silent patient, and he once told me with a certain sympathy, "You are certainly getting a bellyful of resistances." But at the end of the year he told me that since she was invariably silent, he could not count the forty sessions we had together (four thousand dollars in his pocket) as credit toward my graduation, and I had better find a new patient and begin my supervision with him all over again. I could, however, keep the dollar fee she was paying me.

But while I did not learn much from my supervisor on this case, I was learning a great deal from my patient, not about her, but about myself. I was in an agony of boredom: not a religious man, I found myself praying that she would miss a session. Not a chance. I was willing to remain content with her simply being a few minutes late. No such luck. She was *always* punctual, and when once, after forty-eight minutes of total silence, I said our time was up, she turned with a triumphant smile and said she believed we still had two more minutes of the session to go. No amount of interpretation ever got her to examine her glee, however—and if she thought she was boring me to death, clearly she would not have been satisfied until she heard the definitive thud behind her.

She was lying on a couch in front of me, and could not see me. This became a small comfort to me, as I found myself fighting an increasingly strong desire to go to sleep during the session. I would be lying if I said I never gave in. I did. Frequently. The reason I stopped was not because my colleagues so disapproved, though that was also true. (I did try to discuss it once in class, but when I confessed to falling asleep in a session none of the other candidates would admit that it had ever happened to them, and so I wound up looking foolish and ill suited to the profession, which by now I was beginning to suspect on a daily basis in any case.) The reason I exerted myself to stay awake was because I was once aroused from a deep slumber by my silent patient engaging in one of her monthly forays into speech. "Dr. Masson, you have been sleeping." Too stunned to think, I went on automatic analytic pilot. "What gives you that fantasy?" "Because you were snoring," was the unanswerable retort.

As soon as your first supervisor indicated that you were making progress with your first analytic patient, you were allowed to begin seeing a second patient. The rule (again, unexplained) was that it had to be a male suffering from an obsessional-compulsive disorder. I was clearly not making progress with my first patient; nevertheless, about a year and a half after I started seeing her I was given permission to begin my second case. Possibly one of the training analysts had a free hour and wanted to fill it with a paid supervision.

It was not difficult to find patients for these supervised analyses, because the fee was so low, between one and ten dollars a session. Sometimes patients were sent by the psychoanalytic "clinic," and sometimes people were referred to us by our supervisors, or even by another candidate. Since all but one of the candidates were psychiatrists, they had access to patients continually, either through their hospital affiliation or through their private practices, which they all had. We were told not to reveal to the patients that we were merely candidates rather than full-fledged analysts, since this would have an adverse effect on their expectations, but the referring psychiatrist or psychoanalyst generally had already made this clear to the prospective patient. (Ironically, we had to sign a paper upon admission to the program that we would not call ourselves analysts until given permission by the institute to do so.) For many candidates, the patient's initial disdain often proved a difficult obstacle to overcome. In retrospect, it was unwarranted: I believe the patient was as likely to get bad treatment from one of the senior analysts as from a candidate. And at least the candidates were supervised, so egregiously outrageous behavior was less likely.

I was now ready to take on a second patient. I told a number of psychiatrists I knew that I was looking for a second clinic case (so called because they paid only clinic fees), and after a few weeks I was told by the university psychiatric clinic that they were sending somebody over to see me.

My second patient walked into my office and said, "Remember me?" I did, actually. He was an undergraduate student in one of the literature survey courses I had taught. I remembered him as very quiet. I was not sure about the propriety of seeing a patient who had been one of my students. As it turned out, however, my supervisor did not think this was a problem, since I had hardly interacted with him, and presumably he knew nothing about me personally. He gave me a brief outline of his life, which sounded extremely constricted. He had grown up in a Ukrainian neighborhood, attended Ukrainian schools, had Ukrainian friends, and his ambitions, it seemed, were limited to becoming a Ukrainian father with a Ukrainian wife and Ukrainian children. I was not happy to hear this. My first case was extremely problematic, and now my second promised to be boring. How would I be able to feel empathy with his life? And if I could not, how could I possibly help him in analysis?

I was wrong. Ivan* turned out to be an ideal patient, at

least from the psychoanalytic perspective. His dreams were vivid, fascinating, and easy to interpret. I can well remember the first dream he told me. "In the dream I left my house and walked to the corner. There is something beyond the corner that I very much want to see, but I cannot seem to turn the corner. I go back to my house, and set out again, and again when I reach the corner, I cannot get beyond it. That is all." That was enough. The dream was so simple, so classic, I wanted to shout my interpretation. Clearly he was trying to get beyond the confines of his own ethnicity set for him by a highly restrictive mother, but could not. Here was something I could genuinely help him with, because I understood this wish, and had sympathy with it. But he did not see that this was what he "wanted." The wish to be free and independent, away from the confining atmosphere of his family, was actually unconscious. The fact that it existed, and was nevertheless outside the realm of his consciousness, gave me great respect for Freud; here was something Freud wrote about constantly, and I was actually witnessing it first hand. We want something, but cannot or will not acknowledge that we want it. Still, I reasoned, if Ivan could not admit to himself his wish, but had to disguise it in the plastic form of a dream, then it would not do to simply blurt out the answer. That much I knew. But beyond that, I was at a loss. How was I to proceed?

I was fortunate in that my supervisor for this case was Peter Thompson, the benign analyst who had interviewed me some years before and who had taught my first seminar. He was genuinely helpful. By nature a patient man, he put a brake on my own impetuosity. It is not that he understood this man any better than I or the man's own friends did, but he was anxious to see that I not ruin any chance I had of helping him by rushing in with my supposed greater understanding and forcing him to confront something prematurely. Still, I thought if Ivan's dreams were so plain, the knowledge could not be so deeply buried, and probably he was ready to turn the corner quite on his own.

He was. He just needed someone to listen to him. And listen I did for the next three years. It was not bad for me, and it was not bad for him. While he made "progress" and became much more independent, he was at an age where I think he would have made this same progress quite on his own. I cannot really take credit, though I am sure I did not put any further obstacles in his path. I have never entirely understood the strange paradox of somebody who was such an ideal patient (he continued to have fascinating dreams) yet would have made, I imagine, such a poor friend for the analyst. It was entirely in contrast with my next patient.

But before I come to my third case, I should say something about the scientific meetings, as they were hopefully called. Candidates were only allowed to attend these august occasions, which were held once a month, after we had begun our second case under supervision, since this would be the first time all of us would come face to face with our analysts in a nonanalytic situation. We would hear them present cases, argue with other analysts, and watch them behave like the human beings they presumably were, as opposed to the gods they were in their own consulting rooms. We were all at a fever pitch before the first meeting, anticipating how our analysts would behave, the great secrets we would be let into, and the high level of scientific discourse that reigned at these confidential meetings.

About twenty analysts were present for the first scientific meeting attended by the candidates, held in the boardroom of the Clarke Institute of Psychiatry. The atmosphere was severe. People nodded to one another, and some of the senior analysts, clustered into small groups, spoke among themselves in soft voices. I do not think most of the candidates knew how much animosity existed among the different groups, and how many analysts were no longer speaking with their colleagues. Schiffer made me privy to this gossip. The candidates huddled together, too stricken with awe and fear to say a word. We sipped our weak coffees with trepidation, as the meeting was called to order. There were no pleasantries. The chairman announced that Dr. Brian Bergman would present a paper on narcissistic grandiosity in a female patient. My friend Mike Allen leaned over to me and whispered, "He ought to know." None of the analysts in Toronto had published very much, but Bergman was exceptional in having published virtually nothing about psychoanalysis. I kept trying to catch Schiffer's eye, and I imagined he looked back at me several times with a meaningful look, as if to say, "What shit, eh, Masson?" Bergman droned on and on. He spoke about a woman who could not "own" her emotions. It is not that the topic was intrinsically uninteresting—it wasn't, and the woman sounded fascinating. That was the problem. She was so much more colorful than the dreary Brian Bergman, who struggled, unsuccessfully, to bring her to life. Suddenly I had the feeling that this woman had never given me permission to hear about her life. Nor, I am sure, did she give Bergman permission to relate it with so little art. She probably had no idea to what use her life story was being put this evening. Her privacy was being invaded by this scientific meeting as surely as if we were a gang of voyeurs. And is this not what most of us were, and rather bored voyeurs at that? The paper was dreary. The meeting was dreary. The evening was dreary. Is this what I had waited for for so many years?

The candidate's third case was supposed to be a more "serious" one than the hysteric and the obsessive-compulsive neurotic. The person was supposed to be suffering from a "character disorder," that is, whatever was wrong was more than a neurosis but less than a psychosis. Something fundamental was askew, or flawed. To me, looking at other people in terms of what is wrong with them—these gradations of disturbance— was and is distasteful. Always implicit in the doctor's view is, of course, how much more "healthy" you are than they. And this is almost never the case.

The woman who came to my office and was to be my third

patient was in all respects the opposite of my second patient. She was an attractive, mature woman, with grown children. "My father was a bigamist," she told me at our first session, "and had two families. We didn't know for years." She lived an intense life; she was passionate, held strong opinions, had friends she loved and enemies she hated. I liked her immediately. Here was somebody I would be happy to have as a friend. But, mystery of mysteries, her analysis was anything but a success. Unlike those of my second patient, her dreams were never transparent, at least to me. I was delighted to listen to her, but had no clue how I could help her. Her "symptoms" were a general feeling that her life was not going well, that she was not as happy as she could be or should be. She was deemed to be suffering from an "inadequate personality," a diagnosis based primarily on her having made a disastrous marriage (who hasn't?) and sacrificed a great deal for her children. She was often unhappy, although I always saw a good reason. I think the diagnosis was applied to her because members of the institute learned that she had "rejected" a therapist, thereby demonstrating her "inadequate personality." Today, the *Diagnostic and Statistical Manual of Mental Disorders*, (DSM-IIIR), the official diagnostic guide of the American Psychiatric Association, used by all psychiatrists, and heavily relied upon by insurance companies, holds that for a patient to state that a therapist is boring or uninteresting is a primary sign of what is now called "the self-defeating personality disorder."

I did not expect that she would develop a strong and intractable eroticized transference, but she did. Try as I might, I could never budge her from the firm belief that she was in love with me. Her analytic progress was in inverse proportion to that of her external life. She did fine, things went well for her, but we got nowhere as doctor and patient. To make matters worse, whereas in the case of my second patient I had had a benign supervision, in this case I had an unsympathetic one.

Dr. Stan Lesser was a particularly unpleasant man. He spoke almost inaudibly in a kind of mumble, and he chain-smoked throughout our sessions. When he laughed for no particular reason, which was frequently, the laugh was mirthless, revealing tiny little brown teeth. Watching him was distasteful. Listening to him was no delight either. His favorite phrase was "Why did you say that?" But I came to understand that he did not expect an answer, it was merely a statement. Soon I dreaded my supervision with him. I think the feeling was mutual, because he started to show up at least half an hour late for the sessions, and ended them well before the hour was up, so that in the end I was paying him a hundred dollars for about ten minutes. I ranted and raved to friends, but one of them pointed out to me that I should actually be grateful, since I had saved forty minutes of unpleasantness.

Bad as the sessions were with Dr. Lesser, they were not the worst. The worst were reserved for the last supervisor I had. The time for me to graduate was approaching. I was beginning what I hoped was the final downhill run to graduation from the institute, to becoming a psychoanalyst—a member of an elite profession and legally entitled to practice psychoanalysis, from which I could reasonably expect to make a very good living indeed without renouncing my position at the University. In 1975 I had already been made a full professor. I had only one more patient whose analysis needed to be supervised. All of the supervisors, however, were fully booked. Nobody had any free time, nor would they have any for a year. I did not want to wait another year to graduate. There was only one thing I could do.

A new training analyst had moved to a town in Ontario, some four hours' drive from Toronto. Maybe he could squeeze me in. I called. He could see me, but only, unfortunately, at eight A.M. on Saturdays. Did I mind getting up at four A.M. to be at his house every Saturday morning at eight A.M.? No, I didn't mind, especially given the driving conditions in the Canadian winter. What was four hours of blizzard conditions if at the end of it I could be supervised on my last case?

I was wrong, I did mind, but not so much the drive. Terri

selflessly offered to go with me, and we had some of the best talks of our marriage on those long, subfreezing treacherous drives through the snow-blanketed Ontario countryside. What I resented was arriving there at eight A.M., to have Dr. Blaustein* look puzzled to see me, yawn sleepily, glance at his watch, and ask if I minded waiting while he had breakfast, since it was so early. I minded, I minded. I was famished after the long drive and I would have been grateful for a cup of coffee and some toast, but it was rarely offered. This was, I realized, all part of the hazing that goes with seeking admission to any fraternity. The problem was, I hated fraternities and disapproved of hazing. An army bootcamp, fraternities, the priesthood, all of these semisecret societies depend upon a certain degree of humiliation in order to acculturate the candidate into the mad world he has committed himself to.

It is strange how certain unpleasant experiences have a way of resonating for years afterward. I can clearly remember the feeling of arriving in the sleepy, snowbound provincial town early on Saturday morning and wondering what I was doing there. Was anything worth that drive, every single week for two years? In a way, yes. Just seeing the town; the silence of the snow; the autumn leaves falling; the houses lighting up for breakfast on those cold, dark mornings; and the feelings the atmosphere evoked (captured in print by the fine Canadian novelist, Margaret Lawrence)—feelings of loneliness, of desolation, of isolation. For me, being able to have the feelings, then leave the town, became almost an internalized metaphor of escape. It wasn't forever, it could be overcome.

Dr. Blaustein had been trained abroad and was a great admirer of the British "object-relations" school (which placed even greater emphasis than did Freud on very early relations with significant "objects," i.e., people), especially Harry Guntrip and Douglas Fairbairn. He was also much influenced by the teaching of Michael Balint, one of Sandor Ferenczi's pupils, and of Melanie Klein, a child analyst who had been the greatest rival of Anna Freud. When I first told him about the cold, bleak childhood of my patient, he cut me short by telling me that it seemed to him she had "good enough mothering." I was struck by the ordinariness of this term, and he patiently explained to me that it was a technical term used by Donald Winnicott to refer to—and here he hesitated—to, well, mothering that was good enough. Perhaps Winnicott developed this notion in the face of Melanie Klein's notorious defects as a mother. She asked Winnicott to analyze her son—which he did, though not under her supervision, as she demanded. I thought of the long drive back, still ahead of me in what promised to be a nasty snowstorm. I looked around his large, warm, and comfortable home. It clearly gave him great pleasure. Didn't he feel the absurdity of sitting here and fulfilling this final ritual when both of us would rather be doing something else?

I admit, though, that I did want somebody to talk to about my patient. Daniella* was a young, attractive graduate student, born in Eastern Europe. She was engaged to be married and had been in treatment in the university clinic with a woman psychiatrist because she was not sure she loved her fiancé, her life felt empty, she was often unbearably anxious for no apparent reason, and because she was tormented with strange dreams. When she was small, her mother had suffered a "nervous breakdown" and had been hospitalized, and she feared that she, too, would have hallucinations like those of her mother, or at least was capable of them. She had hardly known her father, who had died when she was very young.

Within a week Daniella contrasted her sessions with the woman psychiatrist, where "we just talked nicely," with all the terrible things that were being said in my room.

"She never pried, the way you do. She would ask questions, like 'How is your sexual relation with Michael?' and when I said 'Fine,' that was it."

One time she said, "Last night I dreamt I was on a raft, and there was a man there with me. It was not my choice to be in that rough river, but the man kept telling me to go on, right into the rough waters. I guess you know who the man was."

She described a deprived and lonely childhood. She dreamed of curling up inside me, as a small baby. I was, she said, like a mother and father to her. "You are doing more now than they ever did. My mother never talked to me about personal things. I was so confused as a kid, and nobody ever sat down and told me anything." She would say these things with great emotion, and would describe odd sensations in the office. "I feel the weight again, a big, heavy weight on top of me. It feels like another body. My heart is really pounding now. I am afraid that I am going to scream." She put her hands on her cheeks and began to cry.

"Last night I dreamt that I was in deep, dark water, about to fall off something. I was about to panic when I saw the shore. I swam to it, and then could swim back again, without fear. I know it's all about you."

"I say things in here I would never say anywhere else. How can I get up off the couch and look at you? What you must be thinking!"

"You probably think I am just a little kid. Well, I'm not. I'm a young woman, and I know I'm attractive. And you're not my father. You are Dr. Masson. And you are within reach, sitting right behind me, just a little older than I am. And you have sexual organs. You are too close. You could never be my father, but you could be my lover."

I needed advice—kindly advice—about how to call upon whatever reserves of empathy I was supposed to have acquired by then. But the advice I got from Dr. Blaustein was not intelligible to me. One of the first things he told me, something he repeated often, was "If you feel something in a session, stop and consider it as a possible symptom of the patient." In other words, if I felt bored, or restless, or disgusted, or angry, I should consider that these feelings did not come from me, they came from her, and that I should turn my gaze away from myself. *I* was not to blame for what I felt, *she* was to blame. But in fact I was a separate person, and brought to each and every session my own bad moods, my own problems with Terri and others.

I was as much to blame for what went on there as she was, probably more so. But she was the victim, and she got blamed. On the other hand, when he told me that "the universe she is taking you into is a paranoid universe. You must float along with that paranoia. Do not seek to stop it or even to understand it or you will break the spell," I thought he might have a point, but it was not something I could do. "You want her to remember," he once told me, "and that is your problem, your symptom, your countertransference. It would be better if you remembered less about her. In fact, try not to remember anything about her, not even her name." I understood what he intended, but it felt too much like an artificial gimmick. There was something admirable in meeting every person with a completely unbiased mind (unbiased even by a memory of who they were), but it was not credible to me. I could not pretend I was not interested in understanding, I was. Nevertheless, the primacy that she gave to feelings, and his insistence that I should try to enter such a world, no matter how difficult for me, led me to reflect that I had my own problems with experiencing emotions without the mediation of thought. This last patient was testing me in a new way. I was beginning to feel that I could not help her because I lacked some gift, some capacity for emotional empathy that others I knew seemed to have.

Meanwhile, Dr. Blaustein was trying to persuade me to accept the kind of theory that was found in the writings of Melanie Klein. I had read enough of her work to know that I abhorred it. But now all of her words—the good penis and the good object, the bad breast, introjecting the hostile and the bad internal object, the bad faeces, hostile projective and introjective identification, persecutory anxiety, the depressive position, and the paranoid position—were being hurled at me till I felt like the baby being forced to drink poison milk (one of her favorite interpretations), or like the young boy-patient who could no longer bear to hear her insistent and repetitive interpretations and fled a session in misery saying he wished he "were not there." She had a relentless disregard for the particulars of human experience—where something happened, what language was spoken—and a fascination for delving into remote areas that seemed beyond speech. Forcing Melanie Klein's ideas on me was of no use to me, and it just made the cold world outside seem all the colder that I could never establish any kind of human bond with this man, my last supervisor on my last case. But I went, every week, faithfully, for two years, until one day for obscure reasons I could never understand, he told me I could now continue on my own.

The rooms at the Clarke Institute of Psychiatry, where all our seminars were held, were bleak and institutional. The seminars were always held at night, from eight to eleven, to allow time for the candidates to finish their day's work and have dinner with their families. Every single candidate was married—I believe it may even have been a prerequisite, perhaps as a sign of emotional maturity or of social conformity. Coming in out of the cold snowy nights I always hoped that the place would be transformed into a living room, with a fire in a fireplace, and good coffee, and that we would sit around until the early hours talking with animation about the love of our life, psychoanalysis. Instead we carried a cup of weak institutional coffee from the cafeteria machine, and we sat around a large white Formica table distrustful of one another.

In effect, the seminars were more of an acculturation, a socializing process, than a genuine learning experience. Each candidate found himself, knowingly or not, defending theoretical positions that, lo and behold, were held by his training analyst. There were Kleinians, and classic Freudians, and those who followed the British object-relations school, and others who were "American oriented," and even a smattering of Lacanians. These theoretical orientations were overlaid and even superseded by old personal hatreds on the part of various training analysts going back years (Goldfarb didn't invite Bergman to his daughter's graduation; Garbin didn't support Kriegel when he was trying to become a training analyst; that kind of thing). We inherited them; that is, the hostility that actually existed among these training analysts was transferred onto the candidates.

The seminars were taught only by the training analysts, which meant that every night some candidate was in agony, since his analyst was teaching. It is a very strange experience to lie on a couch five days a week and tell somebody everything you feel with absolutely no response from him, at least theoretically, only to be in a room with him one day hearing his ideas about psychoanalysis. Used to crumbs, we were now guests at a banquet. Or so it seemed to us in our need. For almost nobody, including me, was this a disillusioning process, though it should have been for everybody.

At a more specific level, what made the seminars so painful to attend was that the analysts who offered them seemed to feel no compelling reason to prepare for them in advance. Invariably the seminars consisted of discussing our readings of Freud and the subsequent literature on whatever problem Freud was speaking about. We were expected to read through the twenty-three volumes of the *Standard Edition*, James Strachey's elegant translation of Freud's complete works, before graduating. Reading Freud was a delight. But I found that discussing Freud with somebody who had reluctantly read him years before, and had little interest in the text or in knowing exactly what Freud meant, was less than rewarding.

In compensation I was doing an immense amount of reading on my own. I had decided to try to make my way through the entire analytic literature available in English. So I would check out of the library the issues of the standard analytic journals, *The Journal of the American Psychoanalytic Association, The Psychoanalytic Quarterly,* and *The International Journal of Psycho-Analysis,* and I would read through them systematically, starting with volume one, up to the present. While the theory was generally of little interest (apart from those by Freud, the number of really worthwhile articles was very small, perhaps a few dozen), I was fascinated by the case histories. Had somebody selected and presented the truly valuable articles to me at the beginning, I could have read all of them in a few days. The actual ideas I had heard in the six years of seminars could similarly be reduced to a few dozen and could have been conveyed in a few short conversations: the unconscious, repression, the importance of childhood events, the significance of trauma, counterphobia, the repetition compulsion, unconscious affects, the nature of dreaming, etc.

When it came to psychoanalytic technique, reading was helpful but could not substitute for experience. It was in the latter area that I stood my best chance of actually learning something from the seminars. There, I learned a fair amount about what we called the "tricks of the trade," things you did to encourage certain kinds of reactions. For example, one of the members of our class was, by universal consensus, remarkably obtuse. He spoke slowly and thought even more slowly. In the end, what he came up with was usually wrong. He often lamented that he was in the wrong profession. As luck would have it, his first patient was an articulate, very bright graduate student at the university. He complained to the class that whenever he said anything, which was not often, she would go silent, and get the puzzled look of somebody who suddenly discovered they were in the wrong meeting. The advice was easy: he was to say nothing, beyond questioning grunts. After all, had Freud not explained that the analyst was only a mirror? Keep the field clean, don't contaminate it with needless interpretations. Let the patient do the work. He was cheerful a few weeks later, saying that his absolute silence was paying off. "Nobody ever called me profound before," he said admiringly of his patient.

Making interpretations not only risked letting the cat out of the bag as far as our own intelligence went, it also risked humiliation. "Stick to the classical interpretations," we were warned, for they could be memorized from the literature. It was easy to invest heavily in a silly interpretation, only to have it refused outright or ridiculed by a patient. You then had the choice of attempting to persuade the patient, or swallowing your pride and admitting you were off base. Not every analyst had the stomach for this kind of honesty. My own analyst, toward the end of the analysis, once told me, "All your life you have wanted to be . . ."—there was a pause, and I prepared to hear the great secret that was about to be revealed to me, the meaning of my very existence—"a beautiful woman," he finished. He leaned back in his chair, and I could feel his pride. He had played one of his great trump cards. "That is total nonsense," I shot back, because it was, and as soon as I said it, I regretted it, because I knew that he knew I was right. I had hurt his feelings. He was so proud of this interpretation, it must have struck him as very deep and quite original. He shrugged, and I knew that he would try again tomorrow. That was the nature of analysis. It taught me a valuable lesson: do not interpret just for the fun of it, or on the off-chance that you might be right. But without an interpretation per session, the patient would feel cheated. Without an interpretation, of what use was the analyst?

We learned, too, that no matter what the question, when it was embarrassing, or one you did not want to answer, the simple device was to throw it back to the patient: "Why do you ask?" Or, a step deeper; "Do you not see the connection with questions you asked as a little girl?" The compliant patient would usually go on to free-associate to some event in childhood and the situation would be saved. Above all, we were taught never to reveal anything about ourselves. To me, this seemed excellent advice: had the patients known who we were in reality, any of us, they would not have wanted to remain. The trick was to keep them from finding out.

Some analysts went so far as to suggest that the room be as spartan and as bare of the personality of the analyst as possible. Even the journals we chose to display should be bland: I was told not to leave out copies of the *New York Review of Books* or *The Nation* because this would give away too much. We should not state whether we were married, whether we had children, how old we were, what our religion or politics were. The only information that was "legitimate" was about our training, and even here, if they asked what kind of psychoanalysts we were, we should only inquire why they wanted to know, but not answer. We were to become blank screens upon which the patients could project their own fantasies. The rationale was that all of this desire to know was grist for the analytic mill. We did not want to cut off fantasies with knowledge. We had to allow their imagination full play. Knowing would defeat the purpose.

But of course we were not blank screens, in or out of our offices, though many did their best to become like them. And in reality patients learned as much about us from what we didn't say as from the many lapses and the innumerable clues our dress, demeanor, office, reading material, and even the emotional quality of our silences provided. It was something of an eerie feeling to have said so little and to feel so exposed. I think patients had an intense desire to know about us out of a natural self-protective tendency, and there was a grapevine, for patients talked about their present and former analysts. I never accepted the piece of psychoanalytic wisdom that states that every patient gets the analyst he or she deserves. I saw too many cases where kindly, pleasant, affable, and intelligent people linked up, through no fault of their own, with analysts who were cold, remote, unpleasant, and stupid. Other analysts were cruel, even vicious. They too had fine patients. The system seemed devised to hide these defects from patients. It was almost impossible to know, except over the years, what your analyst was really like as a person. Almost every candidate will confirm that it is an odd experience to have lunch with your former analyst, and not merely because of the highly intimate and embarrassing things that were said during the years of analysis. To nobody else, it would seem, do we reveal the most intensely shameful experiences of our lives. But every minor revelation about the analyst comes as a complete surprise ("So he drinks wine") simply because these details of behavior were formerly hidden. For many, it can deal a crushing blow to the idealization. Conversely, the analyst has only a very partial knowledge of any patient, even within analytic terms. To meet the same patient on the outside is often a shock. I had no idea of simple social skills that a patient possessed. When I accidentally met one of my women patients at a play, I was surprised to see how charming she was. This was something that simply never appeared in the analysis. To know so much about somebody at one level and so little at another is disconcerting.

I knew many Jewish patients in analysis with one training analyst whom others considered an anti-Semite. Did it matter? Of course it did, but you could still find today many analysts who would defend this as being a character quirk and not necessarily detrimental to the practice of psychotherapy. I know of an occasion, however, when an analyst's insensitivity to the concerns of a Jewish patient seemed directly linked to her death.

One of my fellow candidates was Catholic, and was preoccupied with Catholic theology. He had the misfortune to have as a patient a Jewish survivor of the concentration camps. During one of the case seminars he explained to the class that this patient suspected him of anti-Semitism. Since he had once complained to me that Freud was too preoccupied with "Jewish" themes, I was sympathetic to her concerns. "I am asking for help," he said, looking miserable. I thought this only fair. No doubt he wanted to ask somebody else, somebody more sensitive to these issues, to take over the case. "How can I get her to understand that this is merely a projection, and a paranoid one at that? She is being chased all right, but her tormentors, her persecutors, are inside her own head. She can't see that, and she thinks her worst problem is that she has fallen on a bad analyst." She was right, I thought. The class and the supervisor all urged him to redouble his efforts to provide this woman with "insight." But from class to class, things got worse. "She is convinced that she is locked into a life-and-death struggle with me, and if she cannot get me to change, she is going to kill herself. How do I get her to see that the change must be in her, not me?" I could not see how this attitude could possibly help her. One day he came to class and was crying. "She killed herself." I think it was one of the first moments that I realized we were engaged in an activity that was no game, but something where our own inadequacies could have deadly consequences. His role in her death was never discussed.

Since Toronto was such a melting pot, many of the patients came from Europe and had lived through the Second World War. I privately wondered how we young analysts-to-be, who had no experience of war and who had in fact only minimal knowledge of what had happened in Europe, could possibly understand the experiences that many of our older patients had gone through. It soon became clear that we could not. This was a greater problem for me than for the others, because they salved their consciences by rigidly maintaining the doctrine that external reality counted for little. It did not matter what had really happened, only how it had been perceived. The language of the person did not matter—the country, the historical and political circumstances, all were irrelevant. This saved the analyst considerable stress, but it could not possibly, I slowly began to suspect, work to the advantage of the patient. We were given case histories by the English analyst Melanie Klein, in which it was never even clear where the analysis had taken place, in war-torn Germany or in postwar Britain. And it did not matter, said Melanie Klein, all that mattered were the good and bad "imagoes" within the patient.

Throughout the seminars it was continually drummed into our heads that external reality was not as important as internal reality. The issue was highlighted whenever one of us would recount the narrative of a patient who had been sexually abused or in some other way traumatized in childhood. The very first time one of us used the phrase "sexually abused," the instructor, Dr. Garbin, one of the more sophisticated of the analysts, corrected him. "*Feels* she was abused," he explained. He went on, "We have no way of knowing what really happened to this woman. We have only her version of the event. You must not buy into it. She could be paranoid, she could be mistaken, she could be hiding her own complicity. You cannot know the reality. But this does not mean that you should accept her version. Besides, if you do, you may come to regret it later, for she may herself change her perceptions, and you might be left holding the earlier version in your own head, unable to free yourself from it." It sounded *sort of* plausible, but it still flew in the face of our everyday experience.

A woman told one of the candidates that she *remembered* being sexually abused by her stepfather when she was thirteen years old. Was her memory not to be trusted? Dr. Garbin began an elaborate explanation of pre-oedipal masturbatory fantasies, which centered on the parent of the opposite sex. When they surfaced in adolescence, these memories of fantasies were disguised as memories of real events. So as to hide the shame and the guilt over those early activities, the masturbatory activities were remembered as seductions on the part of the parent of the opposite sex. The event was reversed, and the person who was passively used in the fantasy became the active aggressor. "Accusations of sexual assault," he explained, "are not real memories. They are screen memories, whose purpose is to hide the shameful fantasies of the past." This was straight out of Freud, I later realized, and had no basis whatsoever in so-called direct clinical experience. But our instructors were absolutely wedded to this opinion, so much so that it ceased to be an opinion and took on the form of a dogma. It became part of the psychoanalytic catechism, and had to be repeated on a daily basis and applied widely to all kinds of situations.

One of the most respected instructors told us that we should all read Truman Capote's "nonfiction novel" *In Cold Blood*. It was constructed, he explained, in the manner of an ideal analysis. Capote was not present in the book at all. He was there, of course, since he wrote it, but he was invisible. The word "I" to indicate the narrator never occurred. This was how the analyst had to be, completely self-effacing. It was an interesting idea, and I went home and read the book. Indeed, Capote boasted in the book that he had a special rapport with mass murderers because they quickly came to realize that he was not passing any judgment on them, that he had no opinion about them regarding the fact that they had killed or committed some other crime (a dubious and, to me, not exactly admirable statement he made in conversations with author Lawrence Grobel).

Moreover, it was naive, of course, to believe that authors can simply convey facts objectively with no interpretation, no sense of where they stand. After all, in Capote's book, which achieved immense fame precisely because it was supposed to tell all the gory details, there is a curious kind of prudery or reticence when it comes to the one salient fact that might have revealed something of the origin of Richard Hickock's murderous sprees. In his confession, Hickock writes, "During my employment there was the beginning of some of the lowest things I have ever done." Capote comments, "Here Hickock revealed his pedophiliac tendencies, and after describing several sample experiences, wrote 'I know it is wrong. But at the time I never give any thought to whether it is right or wrong.' " Capote thus skipped over the actual experiences that Hickock wrote about. In other words, Capote left out of this book that "tells all" what are undoubtedly some of the few pertinent facts about this murderer. Why? I have no desire to speculate, but it cannot be unrelated to something in Capote. So Capote is not just reporting the facts, nor is he without judgment, nor has he omitted himself. He has just tried, successfully or not, to hide.

Can the analyst do any better? I cannot see how. The flesh-and-blood man is there, with all prejudices intact, all the limitations of his own life, all the many things never done, the failures, the joys never tasted. Why should he suddenly be able to transcend these limitations and soar free? While Capote, too, claims his story is true (the subtitle is *A True Account of a Multiple Murder and Its Consequences*), his own biographer, Gerald Clarke, admits that the entire ending was an invention. It simply never happened. Analysts, it seems to me, are indeed like Truman Capote, although I doubt many analysts would agree.

It was with some sense of foreboding that I came to the class given by Dr. Otis Kriegel,* who had just been nominated for training analyst status, on "advanced technique." As it turned out, this class almost cost me my graduation. I didn't like Dr. Kriegel and Dr. Kriegel did not appear to like me. He had recently come from the Menninger Clinic and had the combination of a small-town mentality mixed with enormous arrogance. He was very much the authoritarian psychiatrist who knew best. His advanced technique turned out to be a new system for confronting "resistant" patients: you should employ a social worker to speak to their neighbors, find out the dirt, and then confront them with your newfound knowledge: "That's not what your neighbors say."

I was flabbergasted that this could be presented in a class on psychoanalytic technique. I made my objection mildly, however, by stating that in all the books on analytic technique I had read, including those by Fenichel, Glover, Greenson, and of course Freud himself, I had never seen anything like this.

Kriegel lost his cool. "What the hell do you know? You are a student here; I am the teacher. Just keep your mouth shut and listen."

I rarely lose my temper, but at this rudeness, and after listening to Kriegel's illiterate seminars over several weeks, I could no longer control myself. I jumped up, clenched my fists, and confronted him. "You don't know what you're talking about. You haven't taught us a thing. You have no business teaching at all, you miserable, ignorant fool." He was clearly afraid that I would hit him (I was close to it), and he began retreating, telling me that I needed more analysis, and this was something to take up in my analysis, not in the classroom. He left the room and did not return the next day to teach. He did not teach our class again. I never learned exactly what happened afterward, although I heard that he did everything possible to have me dropped from the program. He did not succeed, but it was apparently close. Kriegel was made a training analyst a few weeks later. My classmates, for the most part, supported me in my version of the events.

My idealization of the profession of psychoanalysis was under a severe strain. My analysis was one of the main reasons. But the seminars, the supervision, my deepening acquaintance with analysts, my getting to know psychiatrists, all this played a role. One of the incidents connected with "the library" was the icing on the cake.

The psychoanalytic library was located in two closets in Dr. Michael Goldfarb's waiting room. It consisted of fewer than a hundred volumes all together, considerably less than my own personal library. Still, I used it from time to time for the journals that were kept there. We checked out the books on an honor system. One day I was checking out the first two volumes of the *Canadian Journal of Psychoanalysis*—in fact, all the volumes there were of this journal, since it never published more than these first two volumes. In these first issues were two brief articles on the history of psychoanalysis in Canada, such as it was, and a friend of mine, a sociologist at the University of Toronto, was writing on the topic and had difficulty in finding copies of the journals. I was borrowing the journals for her. Dr. Goldfarb happened to come out of his office at that moment, and was understandably surprised at seeing me with this journal. I explained that it was really for my friend, a graduate student in sociology. His eyebrows rose. "Oh? I'm afraid we cannot permit the lay public to read these journals. You see, there is clinical material in them, and it would not be appropriate for somebody who is not a medical doctor or in analytic training to see them."

He was not making a heavy-handed joke. He was serious, and I was not permitted to borrow the journals. It was a petty incident, but somehow it signified for me all the narrow, provincial stupidity of the Toronto Psychoanalytic Institute. I knew right then and there that I would never practice psychoanalysis in Toronto, and that as soon as I graduated, I would leave.

Dr. Goldfarb, by the way, was a man who had been described to me as "an analyst to whom you go when other analysts have failed" (though this may only have meant that he needed more patients), who had years and years of experience in listening to people who were in the depths of despair, yet his concerns were focused on a minor journal and the fear of somebody "unauthorized" reading it. It was enough to shake anybody's faith in the profession of psychoanalysis.

What kept me going through all of this was the firm belief that the core of psychoanalysis was good, that it presented a radical new way of understanding the human being, that even if all of Freud's ideas were not right, the method—patiently, humbly listening to another person tell his or her life story—was correct and could be beneficial. I agreed with Freud's earliest theory, that catharsis per se, simply telling one's story, was of value, possibly even "curative." I was also sure that everything I disliked about my training could be understood in terms of the institute in which I was being trained.

While I could genuinely say, now, that the three prongs that make up analytic training were all for me a dismal failure—my own analysis was a disaster, my supervisions were for the most part a painful and expensive waste of time, and the seminars were an intellectual disgrace—yet I refused to believe that what I had experienced had anything to do with "real" psychoanalysis. I was convinced that once I left Toronto and could enter the larger world of international psychoanalysis, all would be better.

Chapter Six

ON ENTERING THE INNER CIRCLE

In 1912 Ernest Jones proposed forming "the committee" of six "loyal" and "conservative" analysts who would guard the royal kingdom of psychoanalysis founded by Freud: Sandor Ferenczi, Ernest Jones, Otto Rank, Hans Sachs, Karl Abraham, and Max Eitingon. Freud was thrilled with the idea, and presented each of the palace guards with an engraved antique Greek jewel from his collection, which they mounted into gold rings and thus became, like Frodo in Tolkien's *Lord of the Ring*, ring bearers. Freud described the committee as a "secret council composed of the best and most trustworthy among our men." He wrote to Jones saying that "this committee would have to be *strictly secret* [his emphasis] in its existence and in its actions."

Before Freud died, Anna Freud received one of the rings from her father. She, too, kept up the idea of a special group of men (she included no other women) who could be counted upon to guard the sacred flame. Their existence was not public knowledge, but it was known that she always maintained a special relation to a powerful male figure. After the death of her father, the mantle was passed to Ernest Jones, and when he died in 1958, to Willi Hoffer, a Viennese analyst living in London. Upon Hoffer's death, in 1967, Kurt Eissler became Anna Freud's closest confidant about matters having to do with the inner workings of psychoanalysis.

For some time I had admired a group of analysts whose

113

scholarship or ideas I found exhilarating: Max Schur and his thoughtful if somewhat tormented *Freud Living and Dying;* Otto Isakower's strange papers on vision; the dazzling papers by Siegfried Bernfeld on Freud's biography (they had been liberally used by Ernest Jones in his three-volume biography); and the quirky, arcane books and brilliant articles of Kurt Eissler, a psychiatrist and analyst, the only one among the four still living and Anna Freud's confidant.

Kurt Eissler was clearly a member of the inner circle of psychoanalysis. Moreover, he was rumored (falsely, as it turned out) to have been close to Anna Freud's father as well. There were many rumors circulating about Eissler, who was called the pope of orthodox analysis. He would give no interviews. He would not allow himself to be photographed. He was a hermit.

What was not a legend was that he had started and maintained the Freud Archives, the greatest repository of Freud's letters in the world, but one whose content was still largely kept secret, and he was the author of many serious and absorbing books about psychoanalysis, published in both German and English, such as the massive two-volume work on Goethe, a book on Leonardo, two on Freud's troubled and gifted student Viktor Tausk, on Julius von Wagner-Jauregg's relationship to analysis, on Freud at the University of Vienna, and still another on lay analysis. His views could be eccentric. At the time, I did not think carefully about what these works actually said in relation to my own beliefs, some of which were very different from Eissler's. But more than anybody else alive, Eissler was knowledgeable about the history of psychoanalysis, and, I later learned, had hundreds of hours of tape recordings of conversations with many of Freud's patients, including a famous early patient, the Wolf Man. If anyone represented the link with Freud and Freud's Vienna, it was the formidable Kurt Eissler.

When I first read Eissler's books, shortly after I entered the institute in 1971, I felt I was entering a world of long ago and far away. It was a feeling that appealed to me, because it recreated the sensations I had had when reading European scholarship during my student years in Paris. I had felt like this when I read the great Buddhist scholar from Belgium, Etienne Lamotte, for example, or the works of the French Indologist Louis Renou. The excitement of seeing genuine scholarship in psychoanalysis provided a link to my own past. I decided to write to Eissler about historical matters in psychoanalysis that had already begun to interest me early in my training. I was surprised and delighted when he answered me and took my questions seriously. I wanted to know about Daniel Paul Schreber,† and more about Wilhelm Fliess, and about Freud's early case histories. I greatly looked forward to getting to know him.

We finally met in 1973, while I was still a candidate in Toronto, at an annual meeting of the American Psychoanalytic Society in Denver, where I was presenting my first analytic paper on "the mad Dr. Schreber." Eissler and I immediately hit it off, although our friendship began on a curious note. I had never seen Dr. Eissler, nor he me. But he was not difficult to recognize. When I caught sight, of a tall, gaunt older man—at the time he was in his late sixties—in the lobby of the hotel where we were staying, looking like someone who had just stepped off the boat from Europe, dressed severely in a black suit with an almost haunted look about him, I knew it was Eissler.

And so I approached him. "Dr. Eissler, I presume. I am Jeff Masson." Eissler was genuinely taken aback.

"How did you know it was me?"

"Well, it was obvious."

"No, no, there is something else. There is something un-

---

†Schreber was a German high-court judge incarcerated in a mental asylum for his "delusional" beliefs that he was being persecuted and that he was the subject of bizarre medical experiments designed to castrate him. Freud thought he had found the source of these delusions in Schreber's unconscious homosexual fixations on his father and his doctor, Paul Flechsig. In fact, his father had subjected him as a child to orthopedic and psychological torture, and Flechsig performed experimental castration on "hysterical" women in the same asylum where Schreber was confined.

canny about this." He did not seem entirely certain that I had not used witchcraft to recognize him.

In many ways it was an unexpected friendship. Eissler was much older, and seemed to be everything I was not: conservative in his dress, brusque and apparently unfriendly in manner, spare in speech. But what Eissler and I experienced together was, while completely nonsexual, nonetheless romantic in some important sense of this word. For one, it was shot through with fantasy. For another, we both behaved as if we were somehow infatuated, both intellectually and emotionally.

Part of the reason was that our beliefs, or in some people's views our prejudices, *seemed* to coincide. I had a great need for loudly proclaiming mine and seeing if I could find anybody who agreed with them. I rarely did. For example, I believed that psychoanalysis was diametrically opposed to all of the major ideas within classical psychiatry. At first I had expected, naively, that other analysts would share my low opinion of psychiatry. It always came as a disappointment to me to hear that a prominent analyst was active in psychiatry, though in fact many were. I found it an even greater shock to learn, for example, that the influential psychoanalyst Edith Jacobson tried to induce a number of her analytic patients to submit to electroshock treatment. I could not imagine a less analytic procedure than electroshock, and I was convinced that other analysts would agree. They did not. When analysts like Margaret Mahler used psychiatric terms such as "predispositional deficiency" to speak about children she considered "autistic" or "psychotic," I was disgusted, but I was convinced I was not alone. She greatly admired Leo Kanner's work in this area, which I abhorred. It was not only Mahler's language that I objected to; she believed, for example, that infants, "with varying degrees of intensity of cathexis, represent a body part for the mother, usually her illusory phallus." I think, now, that I was indeed alone. But I desperately longed to find a kindred spirit, and in Eissler I thought I had found such a person, a man who loved and hated with the same intensity, though less vociferously.

Within minutes of our meeting, I was telling the apparently enthralled Eissler all about my "prejudices" regarding the more arcane realms of psychoanalytic theory—books, papers, and analysts whom I positively loathed. If somebody I met talked with rapture about the "object-relations" theories of Harry Guntrip and Donald Fairbairn, or if they told me how much they loved Melanie Klein's ideas, or spoke with enthusiasm about the works of Margaret Mahler, or enthused over the "French Freud" and Jacques Lacan's incoherent works, I knew we would probably have little in common. My prejudices were wide and catholic: I could hate the "superficial" American ego psychology of a Heinz Hartmann as much as any follower of Jacques Lacan. But I also hated the Lacanians. (This hatred, I hasten to add, was purely intellectual—it was based on ideas, not on people whom I knew.) Terms that were in vogue, like Heinz Hartmann's "average expectable environment," (in the real world, there is no such thing), left me cold. Melanie Klein and her followers, with their tendency to obliterate the real world, alarmed me most of all.

I told Eissler, at that first meeting, that I could remember reading Melanie Klein's *Narrative of a Child Analysis*, an analysis of a ten-year-old boy, Richard, that Melanie Klein conducted during the Second World War. There was a moment in the book when Richard, the boy, told Mrs. Klein that he had had it with her office. "Let's get out of this awful place," he said. Melanie Klein told him that "this awful place represented his inside which he felt to be awful because it was full of dead and angry people and poison." Richard tried to escape to the garden. His analyst followed him. Once outside, "Richard admired the country, the hills and the sunshine." The poor child was tired of hearing about psychoanalysis, and he begged Melanie Klein "not to interpret in the garden." But his analyst was implacable, and "she interpreted in a low voice that he did not want her interpretations because they stood for the bad things she would give him in contrast to the beautiful countryside." Well, yes. When Richard attempted to dig up weeds from the flower beds,

"Mrs. K. interpreted that he was exploring Mrs. K.'s and Mummy's inside and pulling out their babies." Richard picked up stones from between the flowers and threw them angrily against the wall. Who can blame him? Ralph Bion had gone so far as to deny that accurate memories were any more valuable than inaccurate ones, an idea dear to the hearts of psychoanalytically oriented literary critics as well. This was in opposition to everything I believed most fiercely. I liked the "sensible" critique of Klein made by Edward Glover and found his works on technique valuable; the clarity of his writing appealed to me enormously. The book by Ralph Greenson on psychoanalytic technique seemed to me in the same tradition, and hence I liked it, as I did the works of Otto Fenichel, Phyllis Greenacre, Robert Fliess, and to some extent Lawrence Kubie and Bertram Lewin.

Eissler listened to all this with many gestures of agreement, and seemed relieved at the extent to which we seemed to agree, as if he too wanted a friend who shared certain basic assumptions. He, too, had his list of favorite likes and dislikes within psychoanalysis, and it was not long before we had established that we shared many of the same biases. I was especially eloquent on the books and analysts I detested, and Eissler clearly took considerable pleasure in my open and animated manner of displaying this dislike and expanding upon it without reserve. He seemed surprised that our opinions coincided so often. He was not accustomed, in the world of psychoanalytic politics, to meeting people who took no care to conceal their real views. I was a true outsider, and I suspect this trait appealed to Eissler. I had no fear of expressing myself, because I was too young and too junior to hold any office or to be embroiled in any kind of psychoanalytic factional politics. I was consequently free to express my mind. And I did so, to Eissler's apparent joy, over and over.

We sat talking in the coffee shop for four hours. Then we moved into the lobby. I told Eissler my thesis about Schreber, about how Freud was wrong to ignore the writings of the father, and Eissler defended Freud. We were already arguing. But there was a verve to Eissler and a joy in intellectual combat that I had deeply missed in Toronto. He did not look at me suspiciously, wondering about my motivation, he simply plunged into the historical argument. It was exhilarating, and I felt that I was emerging from an intellectual prison. In sympathy with my difficulties in Toronto, Eissler told me one of his favorite episodes in training, Otto Isakower's story of leading a group of psychoanalysts who were teaching reading seminars. One of the teachers said, "I think it is indispensable for the teacher of a reading seminar to have access to the complete records of all the students in his class." Isakower answered by telling the old story of the boy who brings home his report card on which is written, "Murray smells," upon reading which the father writes back, "Dear Teacher, My Murray is no violet. You are not supposed to smell him, you are supposed to learn him!" His iconoclastic spirit appealed to both of us.

Stanley Weiss, one of the organizers of the conference and a pleasant man who had recently moved to Denver from Chicago, saw us sitting together in the lobby and asked us to attend a party he was giving. I wanted to, but Eissler declined. He explained that he did not like social occasions, especially with analysts. We went out for dinner by ourselves, spoke without interruption, and the next day we met again for breakfast to continue our impassioned conversation.

That evening I asked Eissler if he would like to listen to the fifteen-year-old guru, called Maharaj-ji, out of simple curiosity. We went, and on the way back, an odd thing happened. We were riding in a taxicab, and the driver asked me where we had been. I told him, and he began a long, rambling conversation about his son and gurus, and something of his life—and I was actually quite fascinated and asked him a series of questions. Eissler sat in stony silence. When we got out I turned to Eissler and said, "He was an interesting man."

Eissler burst out, "He was not the least bit interesting. You wasted our time. We could have been talking about analysis." Eissler had very little time for small talk, and for him, anything that did not deal directly with psychoanalysis and its history fell into that category. As long as Eissler and I were caught up in purely historical discussions, we were the perfect conversation partners. But I saw that in just about any other matter, our ideas would clash. I did not want to face this fact, because I was getting too much intellectual pleasure from our talks, and I knew that it could be protracted for years. Indeed, in a sense it was. Eissler and I began a single-minded conversation that week in Denver that continued for many years. But I also knew that our friendship thrived in a kind of rarefied atmosphere. When reality intruded, we both found that we had very different values. For Eissler, as I later discovered, nothing, absolutely nothing, mattered more than psychoanalysis. I came to think he would sacrifice anything, and especially anybody, for what he saw as "the cause." Friendships were secondary to analysis. This was like a perpetuation of Freud's attitude. Freud was always talking about *"die Sache,"* literally, "the cause," which was the word he used for psychoanalysis. Everything, for Freud, came after *die Sache.* And so it seemed to be with Eissler.

At the conference, Eissler introduced me to Otto Kernberg, the man who more or less invented the term "borderline personality." I asked him if he was interested in the Schreber case. "Not particularly," he answered. Eissler and I looked at each other as if we had been struck. How could *anybody* not be interested in the Schreber case? In retrospect, Kernberg was probably simply attempting to avoid a passionate and to him possibly boring conversation on Schreber. But afterward, Eissler and I shook our heads in disbelief and analyzed his work in the light of this gap in his curiosity. Eissler asked me what I thought of his writing, clearly hoping I would speak my mind. I did.

"I think the term 'borderline' borders on the obscene, and the ideas behind it are nonsense. He writes horribly, and far too much. Besides which, he strikes me as much more of a psychiatrist than a psychoanalyst. The whole idea of 'diagnosing' is not very analytic. In any event, how could you be interested in so-called border states, and not be interested in Schreber?''

Eissler and I were practically inseparable at that meeting, and from then on, over the next five years of the rest of my analytic training in Toronto, I wrote to him regularly. We sent each other our papers, and I noticed that Eissler almost never wrote something that was trivial. Whether I agreed with his ideas or not, I had to recognize that he was always engaged in serious, controversial, and significant research. When it came to Freud the man, however, there was nothing too trivial for his close scrutiny. I remember once, when we saw a silent film of Freud speaking in the last year of his life to a childhood friend, Eissler wondered if it would not be possible to analyze the movements of the mouth to discover what words Freud was uttering at the time. No word from Freud, written, remembered, or recorded was ever trivial to Eissler. What he felt for Freud seemed to border on worship. At the time, what I felt for Freud was only slightly less than what Eissler felt. I was completely mesmerized by Freud's ability to write with extraordinary clarity. It was a gift that few scientists have, and I loved it passionately. This meant that I could feel complete contempt for lesser minds who criticized Freud without having any of his gifts.

Slowly, over the years, my relationship with Eissler became closer. But Eissler was a very formal man, and remained so throughout our friendship. He never called me anything except Professor Masson, and I never called him anything but Dr. Eissler.

I liked visiting Eissler in his home in New York. His office was a delight to me: completely buried in papers, articles, and books. What mattered most for me and seemingly for Eissler during my visits was that we got to sit in his office and talk psychoanalytic history. It is hard for me now, from this distance, and with all that has happened in between, to recapture the mood it put me in, but there is no doubt that I was completely absorbed. I felt, rightly, that I had a great deal to learn from Eissler, and I was a good and willing pupil. Somehow, too, it seemed "significant," something that had always been lacking in my life when I was a professor of Sanskrit. Eissler and I could move, in one sentence, from some obscure topic in the history of psychoanalysis to the nature of fear. Why, I asked him, did people seem to seek out situations that actually terrified them? "Ah," he would say, "you are asking for the source of counterphobia. Did I ever show you the unpublished discussion of this by Fenichel?" Or when I would talk about my analytic training, and what I considered wrong with it, he would pull out the minutes of the New York Psychoanalytic Society from the early sixties, its heyday, and read with evident pleasure the eccentric comments by Isakower.

When I said I wanted to know more about the Wolf Man, he said he was delighted to hear it, and produced a box of over one hundred tapes. "What is that?" I asked. "My interviews with the Wolf Man. You are welcome to listen to them." Over many years Eissler had met with the Wolf Man, and had recorded all their conversations. He was hoping that someone would come along and sort through the tapes and produce a book. I wanted very much to do it. These were of course not the only tapes he had. He had made a concerted effort to meet every patient of Freud who was still alive, beginning in the early fifties, and he had succeeded to a remarkable degree. He showed me hundreds and hundreds of tapes of these interviews, and eventually let me listen to the ones I chose. It was a veritable treasure trove of psychoanalytic history. Having all these "toys," these distractions to play with, Eissler and I had no need to understand each other, to find out more about who the other person really was. We shared an interest, a hobby, really it was more like an obsession for both of us. Eissler had something I wanted: knowledge. I seemed to have something he wanted: passion. There were many qualities in Eissler that I admired then (and still do now): his scholarship, his generosity, his intelligence, his old-world charm, his integrity, his bluntness, his honesty. But he held one thing sacred, and hence beyond any criticism: Freud. I held nothing sacred and nothing beyond criticism. And with this difference, our friendship was bound to founder.

In Denver, I did not know that meeting Eissler in this way would open many doors to me. I saw it as simply a personal friendship, but as I realized many years later, Eissler was, from the beginning, also interested in finding an "heir." He was seen as remote and unapproachable, yet he had an aura of authority about him that no American analyst shared. He was considered, rightly so, to belong to a special group, access to which was barred to most other analysts. It did not go unnoticed at this conference that I was spending a great deal of time with Eissler, and that he obviously enjoyed my company.

There is a hierarchy within psychoanalysis that revolves around a proximity to Freud. The greatest prestige is reserved for those closest to him. It is not unlike royalty. Everything connected with "his" person is more or less sacred. It was enough, to make an analytic point, to say the equivalent of "*Le roi le veut* (It is the king's wish)," with reference to Freud, to silence an opponent. Everybody, of course, considered his own views to be the closest to those of the master. But only one person still alive had the authority to confer divine acceptance. This person was Anna Freud, Freud's daughter, who was closer to Freud than anybody else, both personally and professionally. Anna Freud was a psychoanalyst in her own right, and one with considerable prestige and stature. She was particularly known for her analyses of children and for the many papers she published in this field. One of her books, *The Ego and the Mechanisms of Defense*, published as early as 1936, has attained the status of an analytic classic.

I certainly did not invent the notion that psychoanalysis has always been rife with politics. Any memoir of the golden period of analytic training in Vienna, during the twenties and the thirties, will describe how the Viennese psychoanalytic world was divided into "insiders" and "outsiders." Anna Freud and Helene Deutsch were probably the two most powerful and most feared women in the circle of insiders, because of their relationship with Freud and also because both had been analyzed by him.

It was inevitable, given my friendship with Kurt Eissler, that I would sooner or later meet Anna Freud. Once the ball was set rolling, I was in demand everywhere in analytic circles—not because I exhibited any inherent worthiness, but simply because I was in with the right people, and I was clearly moving ahead.

Did I want to be an "insider"? Yes and no. I had been raised on the fringe of many cultlike groups, and I had always felt a hunger for a certain kind of closeness that membership in such a group confers, a certain kind of solidarity, for the warmth of belonging somewhere that was not part of the mainstream. At the same time, even as a child I never felt that I, or for that matter my parents, either, genuinely belonged. Our family was vegetarian, long before it was common, and we were also alone in our involvement in Indian mysticism. My parents did not really have a "home," for we moved a great deal, nor even a country: my father was born in France, and my mother outside Vienna, but both lived part of their childhood in what was then known as Palestine and then moved to the United States. They did not identify themselves with America. Growing up I felt both a certain disdain for an ordinary sense of belonging, and a hunger or a nostalgia for it that has never entirely gone away.

It was the same for me within psychoanalysis. And now, slowly, I was beginning to find a certain core group for whom I could feel some admiration. The fact that it was historically connected with Freud gave it the aura of almost mysterious authority that I craved. Eissler would bring me to Anna Freud's house in Hampstead, England, and I would sit there and listen to the gossip, mainly about how one or another officer of the International Psychoanalytical Association was not acceding to Anna Freud's desire to turn her Hampstead Clinic into a fully accredited institute within the International. I was on her side in these battles, but they really did not concern me, and they also had a parochial feel to them. Most of all I enjoyed simply being trusted enough to be present. I felt like a child being permitted to listen to an adult conversation, albeit a passionless one. Not only did Anna Freud herself seem to be without emotions, she also expected others to be so. I don't think it was considered in good form to show too great an enthusiasm or too violent a dislike. Paradoxically, I am certain this is one reason Anna Freud tolerated me. My passions seemed to startle her, but they obviously also intrigued her.

At the same time, being more or less part of an inner circle meant that I made automatic enemies, or at least awakened jealousy and envy in those who, rejected by the circle, felt rightly or wrongly that I did not have as great a claim to this circle as they did. One analyst asked me bluntly, "Why you? Why should you belong to this inner circle, when I cannot? Why should you accompany Eissler on his visits to Anna Freud, when Eissler still does not recognize me? I wrote Anna Freud and asked to see some letters in the house. I was given the brush-off. So what's so special about you?" It was not a question I could answer to his satisfaction, but I could feel the anger with which it was raised, and knew that I was, whether I liked it or not, already embroiled in psychoanalytic politics.

In later years, because of Eissler's patronage, I was well received wherever I went in analytic circles—I was also in demand. For example, the prestigious Group for the Advancement of Psychiatry asked me to several of their meetings. I felt completely bewildered at being there, and I am sure the feeling was mutual. Most of those present chaired departments of psychiatry and knew one another well from previous meetings, always held in some elegant "retreat." The publications that eventuated from these groups were often influential in setting the agenda for psychiatry in the years to come.

I had been invited by Sydney Furst and Mortimer Ostow,

both psychoanalysts, to the group on religion and psychiatry. I found them appealing because they both seemed to have a slightly cynical attitude toward their profession and their colleagues. I thought we had more in common than we actually did. I liked talking to Furst and Ostow at a personal level, but as soon as group discussions began, I knew they were in a different world, one I could not abide. Ostow was sarcastic about cults, and I appreciated his wit, and agreed with his views about Hinduism, for instance, about which I knew quite a bit because of my background. "Jeff," he told me, "I am now convinced that the magic powers claimed by an Indian yoga are only a defense mechanism, a means of retaliating against a brutal external reality by setting up an internal reality that is in stark contrast, where helplessness, for example, is replaced by a feeling of being all-powerful. In other words, for me, magic is just a way of spitting at a bad past."

"I agree totally," I told him, and proceeded to give him some examples from the childhoods of Indian "sages" I knew about. All sages, I concluded, had been badly hurt in childhood, and they were escaping into a world of fantasy where they could compensate (I had just written a paper entitled "Melancholia and the Buddha"). It was fine as fantasy, but more dangerous when it was believed. The truth was, I was interested in the *pathology* of religious concern.

"Jeff," Sydney Furst said, "you have to be careful of your tendency to pathologize *everything*."

At the morning meeting of our group, Ostow began by saying he believed that each of us "acts out in current life a pattern of gratification initially established in early childhood." But then he set my internal alarm bells ringing. "It was not very long before Freud learned that the extravagant stories offered by neurotics as personal history of traumatic significance were actually fantasy created by the distorting effects of a pathologically active instinctual drive." Sure enough, Ostow began to talk about the frontal lobes, a subject on which he had written a great deal. I was chilled when he said that "Malmo studied several patients who had sustained surgical removal of frontal lobe tissue for treatment of mental illness," and the tenor of the talk became even more ominous when he referred to the "full and perceptive discussion of the effects of lobotomy upon personality as given by Freeman and Watts in the recent second edition of their book *Psychosurgery.*" Here I was listening to a psychoanalyst praise a book on *psychosurgery.* How had I come to be here?

In fact, Sydney Furst was right: I was interested then in the pathology of everything, not least in the pathology of just such meetings as the one I was attending. I was having lunch one day at the meeting with one of the younger chairmen of a department of psychiatry. I told him my theory about this.

"In what way are you to be let off the hook?" he asked me. "What about your pathology in attending these meetings? You're not just a cultural anthropologist, studying the curious behavior of the natives, are you?"

"You're right. That's actually an interesting point you raise."

"Uh-huh. But why stop there? As long as we're talking pathology, why not inquire about your pathological interest in psychoanalysis? I mean, it must have some explanation, right?"

"I would have to agree with you."

"In fact, can I ask you why you are so interested in pathology, why you are so critical of everything? Is this not subject to analysis?"

"Fair enough." But I was unable then to take the step that would actually take me *out.* In fact, I had no business being there. I was completely miscast. My luncheon companion was trying to tell me this, and I could not face it, because I was already too comfortable with the companionable lunches, meetings with colleagues, trips to other cities. I still did not want to pay the price of membership, which was that critical faculties had to be focused strictly on what French analysts call "the other" and not on the group doing the analysis.

I was teaching in Berkeley in 1979, but the next year I

returned to Toronto for six months. I was invited to give a "scientific" paper to one of the meetings there. The topic was Freud's abandonment of the seduction theory in the light of the previously unpublished letters between Freud and Fliess, and in particular the case of Emma Eckstein, one of Freud's first analytic patients, and the first woman to practice psychoanalysis. "Can you explain the connection between your own analysis and your interest in Eckstein and Freud?" was the first question. Quiet laughter. Dr. William Stauble was in the chair. To my surprise (and relief) he responded quickly that personal questions of this kind were out of place in a scientific meeting. But it was no use. "Your published paper about Freud and anti-Semitism is just as bad as the paper you gave us today. It is totally useless and all right-thinking people will see the flaws in both immediately. Have you anything to say?" I didn't.

I was more embarrassed than angered, and was glad to leave. Later that evening I went out with Mike Allen and Schiffer and two other candidates of Schiffer. I expected fireworks, but it was a muted meeting, and I was disappointed that after so many years one could face one's nemesis with little more than boredom and impatience to be gone.

Through Eissler I met many of the people who had had some direct connection with Freud, such as Helene Schur, an analyst and the wife of Max Schur, whose book on Freud I so admired; and Marianne Kris, the daughter of Freud's old pediatrician friend, Oscar Rie, with whom he played the card game tarok. She became a distinguished analyst and was married to Ernst Kris, also an analyst, who was influential at Yale (he died in 1957 at the young age of fifty-six). I cannot pretend that intellectual snobbery held no appeal for me. Although there was no official connection between psychoanalysis and the university, I was, then, very much taken with the idea of the university, and I liked the possibility of linking the two realms. But I did not realize at the time how well connected this world of prestigious analysts was outside the university, and just how important a role money played. The analysts I was meeting now invariably lived in expensive homes and socialized with other wealthy analysts. They dressed well, owned fancy cars or boats, and gave elaborate parties. They occupied prestigious posts in departments of psychiatry, hospitals, or their analytic institutes. Many of these analysts were not only preoccupied with their own wealth, but were also concerned with the status of their patients, particularly if these patients happened to be particularly wealthy or famous.

It was usual in the days when psychoanalysis enjoyed enormous prestige for celebrities to seek out leading analysts as their psychiatrists. Thus Marilyn Monroe went to see Ralph Greenson, as he was fond of telling just about anybody who would listen. Why he would boast of it is a puzzle, given her tragic outcome and his none-too-savory role in it. He was one of the many men who did not believe her when she said that she had been sexually abused as a child, though there is no doubt now that she was. (Greenson was fascinated with my work on the Freud/Fliess letters, especially the ones concerning the sexual abuse of children, and I wonder if this was not the result of a bad conscience toward her and other patients whose memories he stigmatized as fantasies.) Monroe was also in analysis with Marianne Kris, who saw her for forty-seven sessions, and then sent her to the Payne-Whitney Clinic in New York. The significance of this for psychoanalysis was that Monroe left a substantial part of her estate to further the work of Anna Freud, whom she had seen briefly for analytic help in 1956 (Anna Freud wrote about her that she was paranoid with schizophrenic traits), and this bequest was undoubtedly achieved through her analysts, who were intimately connected to Anna Freud.

It is not, in fact, uncommon for analysts to solicit, usually through roundabout methods, former patients for money to support analytic projects. Chairs of psychoanalysis in medical schools at various universities have been partially endowed through former patients. This was also the case with the Centenary Fund, named for the centenary, in 1956, of Freud's birth.

Romi Greenson had organized this fund for psychoanalytic research in Los Angeles. The fund's chief donor was Lita Annenberg Hazen, naturally a patient in psychoanalysis.

As an insider, I would hear about these solicitations quite directly. I remember disagreeing with other psychoanalysts over the matter. I felt then, and still do now, that it is an exploitation of the emotional relationship with a patient to solicit money, in whatever form, directly or indirectly. It seems to me that the patient, or ex-patient, is in no position, emotionally speaking, to refuse. Even if the solicitation is subtle or "tactful," I find it wrong and morally distasteful. Many analysts argued that patients who had themselves benefited from psychoanalysis should and could help others to benefit as well. I see nothing wrong with a patient offering money spontaneously. But to ask for it overtly, or even covertly, is to me a coercive act.

I am by no means suggesting that all analysts extract money from their patients, but it would be a rare analyst who does not know about the practice. Has any analyst of note ever objected to this publicly? Has any analyst ever written about it, or raised questions about it in print? I know of none. The only explanation I can think of is that the practice goes back to Freud. Later, when I was working in the Freud Archives, I saw many letters that make it quite clear that Freud expected, even solicited, funds from former patients. Indeed, the *New York Times* of March 6, 1990, quotes from a letter that Freud wrote to a patient, the American psychiatrist Horace Frink, in 1922, in which he attempts to persuade him to leave his wife and marry one of Frink's own wealthy patients. "Your complaint that you cannot grasp your homosexuality implies that you are not yet aware of your phantasy of making me a rich man. If matters turn out all right let us change this imaginary gift into a real contribution to the Psychoanalytic Funds." Of course he was not asking for the money for himself, but for psychoanalysis, and hence he would no doubt defend the practice as ultimately being in the patient's best interest as well. But I could not agree.

In the summer of 1973, I met Victor Calef, an older analyst from San Francisco, when I was in Paris at my first meeting of the International Psychoanalytical Association. He was talking with Ed Weinshel, one of the officers of the association, and when I was introduced to them both, I said that I had just returned from hearing Heinz Kohut speak.

"What did you think of him?" Vic asked.

"I thought he was awful beyond words. Small wonder that he speaks of narcissistic pathology; he was one of the most narcissistic people I have ever listened to. [Analysts are fond of using diagnostic terms as insults in everyday conversation.] God help the person who asked a question. He never stopped his answer. He could talk for hours without pause. And most of what he says is needlessly obscure and pompous."

"Good for you," Calef said. "You know, a couple of years ago, Kohut and I were at an analytic conference together, and found ourselves in the same hotel elevator. Kohut turned to me and said, 'Vic, I have something important to show you, in my room. Please come upstairs.' I told him I was very busy at that moment, but he insisted that it was important. So I went. When we got to his room he threw open the door and invited me to look inside. 'What do you see?' he said, with a broad smile. 'Nothing,' I told him. An empty room. 'Look again. Describe what you see.' I did. 'Well, a bed, a table with some flowers on it, and—' But he stopped me. 'Yes, you have got it. The flowers. You see, Vic, I love beauty so much, that when I travel, the first thing I do is go out and buy flowers to keep in my room.' I waited, but there was no punch line. He had dragged me up to his room to show me what a great aesthetic sense he had. In his eyes it was *important* for me to see *his* sense of beauty. You're right," Calef concluded, "he's a total narcissist."

Weinshel went on to give some stories of his own, then

it was my turn, and within minutes the three of us were talking like old friends. Here, I thought, was the group I had been waiting for —people not afraid to speak their minds. They told me they belonged to a group of like-minded analysts and that I should make a point of meeting these men: Brian Bird from Cleveland, Samuel Lipton from Chicago, Leonard Shengold from New York. I was thrilled at the notion that there was a small band of intellectual rebels within psychoanalysis—so thrilled, in fact, that I momentarily forgot W. H. Auden's quip, that whenever more than two people come together you have the beginnings of gangsterism. I liked the slightly jaded tone they used, the fact that both clearly were successful analysts, and yet felt isolated from the majority of their peers in San Francisco, where they were from. They gave me the feeling that they were not easily taken in, and of course best of all, they seemed ready to accept me. Everything I told them made them give each other a look, as if to say, "He is one of us. We have found a brother." It never occurred to me, until much later, that I might find myself critical of *their* work, once I knew it, or that they would disapprove just as strongly of mine. I just assumed that if we disliked the same things, we were bound to like the same things as well. When Vic Calef told me that he knew well and was a great fan of Siegfried Bernfeld, one of my idols, I knew we were destined to be friends. (In the fifties Bernfeld had the courage to write about the corruption rife in psychoanalytic training, even if he used veiled language, and he had done magnificent and fearless research on Freud's life, refusing to back down when Anna Freud asked him not to publish the results of his work on Freud's cocaine studies.)

They asked me if I had read much of Kohut, and when I said I had, they asked for my criticism. I was only too happy to give it. When I finished, they suggested I write it up for one of the international meetings, and in fact I immediately began working on an article that would later appear in the *International Journal of Psychoanalysis* as "A Critical Examination of the New Narcissism." That evening I also told Vic Calef about my horrendous analysis with Schiffer. He was more than sympathetic, responding with a horror story of his own, his analysis with Emil (Windy) Windholz in San Francisco. This led to an engaging discussion about false mentors, flawed parents, and bad analysts, and how one could benefit nonetheless from a terrible, even a traumatic, experience. Of course one did not seek them out, but there was a way to benefit from the most awful experience. I felt I had found a true friend, and had no inkling that this very friendship would provide me with just such a bad experience that I could learn from.

Chapter Seven

ILLUSIONS

Summer 1978

The older analyst told me, "Look around you. Soon this will be your world and your life. It is a good one."

We were up in the hills of the little bayside resort of Sausalito, near San Francisco, in a beautiful Spanish-style house with a magnificent pool set in a yard overlooking San Francisco Bay. This was the first "analytic" party I had been invited to since my move to Berkeley, where I was a visiting professor of Sanskrit, teaching Indian literature at the University of California. It was a sedate affair, sprinkled with local analytic gossip, and the polite laughter that comes with jokes not really meant to be funny. I wasn't sure how I felt about this community I seemed destined to enter. Was it really a good one, as my analytic mentor was assuring me? It seemed too quiet, the talk, the laughter, too polite, the people, too well dressed. Could I learn to fit into this world?

True, I was now a full-fledged analyst, a member of the International Psychoanalytical Association (the parent organization), as well as a member of the Canadian Psychoanalytic Society and the Toronto Psychoanalytic Society. This meant that I could apply for membership to the San Francisco Psychoanalytic Society, a Freudian institute. I could not become a member of the American Psychoanalytic Association, because at that time they did not accept nonmedical members.

Since our meeting in Paris in 1973, I had been friendly with two senior members of the San Francisco Psychoanalytic Society, Victor Calef and Edward Weinshel. They had

135

encouraged my move. Somewhat to my surprise, I was accepted for membership in the society. I was looking forward to giving my inaugural paper, "The Navel of Neurosis: Trauma, Memory and Denial," the one I had written with my wife, Terri, and which Schiffer had claimed as his.

The paper was about the reality of trauma, especially sexual trauma, and how this reality lay at the heart of emotional unhappiness, which I then called "neurosis," in conformity with the thinking of my colleagues. It contained no new historical information, no new documents, but it was an attempt to argue, at a theoretical level, for the central position that trauma must occupy in any theory of therapy. Terri's section on the Warsaw Ghetto, one of the great "experiments in nature" for laying bare the heart of trauma, was profound not only because she had a remarkable capacity, far superior to mine, for theoretical thinking, but also because it was so personal an issue for her.

The meeting room was austere and uninviting. So were many of the people present, all analysts, almost all of them psychiatrists. Terri, as a nonanalyst, was there on sufferance, as was made clear early in the evening when one of the analysts told her, "We're happy to have you. This once." As I looked out on the unsmiling faces in the room and began my talk, I sensed something intangible in the atmosphere, an "us" against "them," and we were the "them." Only now does it occur to me that it was not, at that moment, personal; the members would have reacted to just about any outsider in this manner. I kept trying to put my finger on what was wrong. Was it that I was not an M.D.? Perhaps. Still, these same people had virtually worshipped Erik Erikson, who had no medical degree. Indeed, the founder of the San Francisco Psychoanalytic Society was the brilliant Viennese nonmedical psychoanalyst, Siegfried Bernfeld. I believe it was more significant that what I was saying was bound to strike many people in the room as offensive. I was taking a strong position on a topic that was not often discussed in psychoanalytic circles: the reality of the sexual abuse of young children. It was by this time acceptable to talk about

physical abuse. But just barely. When C. Henry Kempe, a pediatrician, described a whole syndrome, "the battered child syndrome," in the early 1960's, analysts were slow to acknowledge
its importance, either theoretically or practically. The topic I was
raising was even less welcome.

I was certainly not the first to talk about it. The feminists
were discussing it quite openly, that very year, and a fair amount
of attention outside of psychiatric circles was beginning to be
paid to the subject. But analysts were notoriously insular, and
most of what was written outside the analytic journals was never
read. As a young analyst, I was presumably supposed to stick
to the safe topics, a minor clinical presentation on a phobia or
a fetish, so that the elder analysts in the room could give me
the benefit of their greater clinical wisdom. In any event, when
I finished my paper, three official respondents commented in
turn. Victor Calef went first. "Jeff," he said, "I am disappointed.
I can't hide that. I can't even hide the fact that I am angry with
you. I have read many of your papers with fascination. I listened
to this one with disappointment. It reads like the paper of somebody who has not seen very many patients. Otherwise you
would know how often women do, actually, fantasize sexual
seductions."

"I'm sorry, Vic, I just don't believe it. And in any case, I
don't really know how you would ever go about substantiating
this. How do you decide, clinical experience or not, that somebody's *memory* is really nothing but a fantasy?"

Calef's response was less an argument than an insult.
"You have a lot to learn."

Edward Weinshel was next. "Jeff, Freud abandoned the
seduction theory. You must realize that he would never have
taken such a momentous step without having given it good
thought. His reasons were simple ones: new clinical experience
showed him that he had been wrong. You haven't yet had that
clinical experience. You would do well to listen to Freud."
Again, it was not an argument, it was an appeal to authority.

Norman Reider was even worse. "Terri, why have you

included your own personal experiences in the Warsaw Ghetto? This is in bad form, and irrelevant to the theoretical issues. Such remarks have no place in a scientific paper. Jeff, why have you such negative things to say about Margaret Mahler, who is—I am sure I don't have to remind you—a senior analyst in New York, and well respected by well-respected analysts."

This was not a question, but a slap on the wrist. I tried to defend myself. "Margaret Mahler writes sentences that are not English, and uses a horrible jargon: 'the practicing separation-individuation sub-phase.' I can't bear to read her. You know, in the old days, novelists and poets were interested in psycho-analysis. That is becoming increasingly rare. It may have some-thing to do with the horrible writing in our field, don't you think?" He most emphatically did not agree with me. But the greatest shock came during the discussion when Reider, already old at the time, stood up again, and, with his perpetual cough that made you think you were present at a deathbed scene, said he was intrigued by the number of the letter I had quoted from Freud to Fliess about sexual seduction: the letter number was sixty-nine. He winked. People laughed. It was a tired porno-graphic joke. Terri looked aghast, and asked me if she had heard his comment correctly. Indeed she had. The serious issue of child sexual abuse had just been displaced by a smutty innuendo and nobody seemed to take offense. Was this really a "scientific" meeting? And then it was over.

Only one analyst came up to us afterward, a woman who approached Terri and indicated some sympathy for her situa-tion. I could see Terri was grateful for that minor act of solidarity. I felt like a pariah. I tried to occupy myself with the papers in my briefcase, but there was no getting away from the fact that I was being snubbed, and I was embarrassed. I had been made to feel I had made a social gaffe rather than simply presenting a point of view that was not popular among my colleagues. Terri and I walked out to the parking lot by ourselves, and watched the little groups break up with cheerful goodbyes. We were out in the cold, there could be no doubt about it.

The feeling of isolation was only to increase as we got to know the analytic community better. I made a great effort to get to meet every analyst in the area. I invited analysts to dinner, to lunches. I learned a great deal about the different enmities. It was by no means a monolithic community. Each group felt some enmity for another group. I was already identified with Calef and Weinshel, and by and large that was probably the least offensive of the groups with which I might have aligned myself. Calef, to his credit, did not seem to mind our differences about theory, and I saw him frequently. We agreed that there was a great deal of corruption in the world of analysis, and he let me know that San Francisco was no different.

I remember one evening when Terri and I were having dinner with him and his wife in their home in Pacific Heights. It was a slow summer evening, and the four of us were eating outdoors, on a balcony overlooking the San Francisco Yacht Club, where Vic owned a boat. We were talking about corruption.

"One of our leading training analysts is on a special committee on ethics in analysis," Vic told me. I could tell from his tone that he had something to add, but wanted a little prodding.

"And is he a particularly ethical man?" I asked.

"Sure. He thought his son had married the wrong kind of woman. That kind of woman should have been married to . . . him. So he divorced his son's mother, wooed his son's wife, won her and married her."

"Where is he now?"

"Oh, he's been promoted to chairman of the ethics committee."

"Vic, you're joking, aren't you?"

"I wish I were. What is considered bad is talking about it outside of analytic circles."

Calef was fun. He had been a wrestler in college, and still had a pugnacious look about him. And he loved a good argument. He was smart, he was quick, he was irreverent. He was also very human. He loved and he hated. Especially the latter.

He particularly hated his own analyst, a very smooth Czecho-slovakian analyst by the name of Emil Windholz.

On the other hand, he positively worshipped Siegfried Bernfeld. Bernfeld was one of those mythic figures whose reputation was quite out of proportion with what he had actually produced, but was nevertheless well deserved. The word was that Ernest Jones had taken most of the hard facts for his three-volume Freud biography from Bernfeld, who was much the better scholar. And the few papers that Bernfeld had published about Freud before his death in 1953—"Sigmund Freud, M.D.," or the paper on Freud's use of cocaine—were brilliant examples of scientific investigative journalism before the genre even existed. Calef and I agreed completely in our unqualified admiration. And yet Bernfeld had had a great deal of trouble right here in San Francisco. After founding the psychoanalytic institute, he was then forbidden to vote on the grounds that he was not an M.D. Since the very people who were excluding him were his own analysands, Bernfeld became understandably bitter. Calef and I decided that Bernfeld had not been sufficiently appreciated, and that a volume of papers in his honor should be published. This gave me the occasion to begin a small project of historical dimensions: what, precisely, lay behind his conflicts with his colleagues here?

There was a key document to answer this question, an exchange of letters between Bernfeld and the training committee in the early fifties. I wanted to see those letters, but they proved harder to track down than I thought. Finally I found them in the possession of one of the senior analysts, who had been a member of the training committee at the time. As luck would have it, I was invited to a dinner at his house the next week, and thought I would ask him then to let me use the letters.

The party he gave was typical: the table was beautifully set, the food was excellent, but the conversation seemed wooden. Anna Maenchen, a senior training analyst, and then a very old woman, was present. She had been a close friend of Anna Freud. I turned to her and asked her something about her experience during the war. The topic of "identification with the aggressor" came up in the context of children in concentration camps who tried to find little bits of clothing of their Nazi torturers and wear them. Anna Maenchen used this interesting information to buttress the idea of Anna Freud that these children had identified with the Nazis.

"Well, yes, but that is a rather formal way of putting it," objected Terri. "After all, they had nothing. They were doomed to die, and knew it. It is not surprising that they would cling to something that they sadly hoped could protect them."

"I agree with Terri," I added. "That's not really an identification. That is too strong a word. They are simply trying to survive. Somehow, it is a little bit alarming to use this kind of information for psychoanalytic purposes."

"Yes, it is almost like it is being used against them," concluded Terri.

Shocked silence. Had we really disagreed with Anna Maenchen, and by extension Anna Freud? Anna Maenchen's words made it clear that she was not about to tolerate this kind of nonsense from Terri. How could Terri possibly know, she asked, in so many words?

"Because she was there," I could not help answering.

Later in the evening I asked my host if I could have the letters between the training committee and Bernfeld. The answer was simple. "No." I was puzzled. Why not?

"Those letters are private."

"But Bernfeld is long dead, the letters weren't written to you, yet you have them, don't you?"

"That's different. I was the chairman of the training committee in later years, all the documents came to me, and I found the letters there."

"What would be wrong with my seeing them?"

"You might write about them."

"What's wrong with that?"

"These are sensitive letters. Bernfeld is very critical of his colleagues. The public might misunderstand."

I was sure he meant that the public might understand only too well. And yet here I was thinking of writing some obscure article for an obscure psychoanalytic journal. It was highly unlikely that anybody in the public would ever see it, or even want to.

The dinners that Terri and I gave at our home were hardly any better. A well-known psychiatrist arrived one evening, affable and relaxed. He warned me that finding analytic patients would be extremely difficult. Few analysts had a full practice.

"I am lucky, though," he said, "I have more than I can handle. In fact I have a waiting list of three years. But you had better find some other way to make a living. You simply won't get patients. Not unless somebody helps you."

And that somebody was clearly not going to be he. He told me that he "specialized" in seeing wealthy, famous patients who were also involved in the arts, music, or painting. I could see him straining to stop himself from naming them.

True, it was very difficult to set up an analytic practice. Windholz called me one day and told me that he had an ideal patient, one he could not see for a simple reason.

"And what is that?"

"She has no money."

"How could I see her then?" I naively asked.

"You could see her for free. I cannot afford that; I already have more patients than I can see. But you are just beginning. You have no patients."

One afternoon I arrived at the beautiful home of an analyst who taught a course in psychoanalysis at the University of California at Berkeley. His wife had made an elegant lunch. At one point she was in the kitchen getting something, and I told him that I had read an article he had written on the ego, and found it helpful (a white lie). "Honey, come here, I want you to hear this. Jeff has read my article." Here was a presumably successful analyst reduced to showing his wife his importance. Would I be any better if this were to become my life, producing an occasional article published in an obscure analytic journal?

There were many unsuccessful luncheons and dinners. One evening, Terri and I were invited to the house of Marvin Friedman,* a well-known and well-respected psychoanalyst. I had been to his office and was impressed with the reverse snobbery: it was small, undistinguished, unadorned, in an ordinary medical office building. It was, clearly, a point of pride with him. He was taking seriously Freud's words about making himself a blank screen. He did not want his own taste reflected in his office, for this would only detract from the patients' fantasies. Everything must be projection. Nothing must be real. Friedman carried this to an extreme I had not yet encountered, however, for his house reflected this same bland exterior. Unlike most other successful analysts in the Bay Area, Friedman's house was nowhere near the bay. It was altogether ordinary. The dinner table was set in the same manner. Nothing about the Friedmans was going to be revealed by anything in their environment.

Among the other guests were some of his former patients, all of whom had been candidates and were analysts in their own right. It was a small and elite club. Friedman had considerable cachet in the analytic community because among other things he held a key position in the American Psychoanalytic Association, the powerful professional organization that policed individual institutes in the different states.

Dinner was a strained affair. There were many "in" jokes, none of which Terri and I could follow, since we were not part of the club. People laughed a great deal, and I felt increasingly left out. I decided to try my luck. The Pope had recently died, and this was being talked about. Somebody remarked that his death was obviously of greater import in the Catholic community than in the world outside. This led me to speculate on insularity and ethnocentrism in general. Then I attempted a joke: "Being Pope is like chairing a committee in the A.P.A. It is only of interest to a certain small constituency." Forks stopped midway. Eyebrows arched. Silence.

"You know, don't you, that Marvin holds precisely such

a position?" offered helpfully one of the acolytes. Of course I did. That was what made the remark, in my eyes, but obviously only in mine, a joke. I was trying to say that none of us took all of this very seriously, did we? We took it very seriously, their silence told me. Marvin was least amused of all. He made a cutting remark to me. I knew I had made an unforgivable social gaffe.

There was nothing to do now but forge forward, which I did: "But Marvin, come on, let's be serious. You know that a position like this, in fact, the whole idea of this committee is absurd. After all, the last chairman of your committee, which deals with, among other things, ethical issues, was well known for breaching ethics constantly, if I'm not mistaken."

"You are mistaken."

"No, I assure you. He supervised his own wife, who was not an analyst, on analytic cases. This is strictly against the rules."

"That's nonsense. It's simply untrue."

"I heard it from somebody in a position to know."

"Who told you?"

His tone was not benign. How could I say? The whole thing, which I had thought of as a mild joke, had turned ugly. Clearly I would never be invited back to this man's house. I had just made the enmity of a powerful member of the analytic community.

I was on the outside, looking in. Despite my having come to San Francisco endorsed by Kurt Eissler and Muriel Gardiner and a number of other powerful and highly respected analysts, I just didn't belong. What was I doing wrong? Many things, no doubt. Somebody once told me, "You have to make a mark the way a dog has to piss on a tree." Still, had I held back, had I remained anonymous a little longer, would it have made any difference in the long run? I don't think so, for at the heart of all this was my lack of conviction that psychoanalysis was genuine, and such doubts were an offense. I was giving offense at every turn, even when I attempted not to.

To begin with, as I already mentioned, I was not a medical doctor. The American psychoanalytic community is intimately tied in to American medicine. Almost all analysts in America are physicians and psychiatrists, and the medical profession is layered into a strict hierarchy. Every psychiatrist in a hospital is *chief* of some service, or *head* of some department. The terms "junior" and "senior" are taken very seriously. In psychiatric or medical publications it is not unusual for a "chief" of service to take credit for the work of a "junior" author. In this atmosphere, conformity, a conservative frame of mind, and reverence for the senior men (and they usually are men) are essential. I was irreverent, a nonconformist, and not the least bit conservative.

Meanwhile, my practice was hardly flourishing. I had a few patients in therapy and could see that I would be able, albeit very slowly, to put together an analytic practice. But did I want to? The more I saw of my colleagues, the more I began to see historical research as an attractive alternative. Also, the more I saw of my own practice, the less I felt I or anyone else should continue this activity. I did not like being a therapist.

It was difficult in the Bay Area to find analytic cases. Too many people who graduated from psychiatric residencies and psychoanalytic institutes and even from graduate programs in clinical psychology wanted to move here and practice here. But Kurt Eissler used his influence with various analysts in Berkeley to help me obtain a small practice. Within a few months I was seeing two people in psychoanalytic therapy.

My first patient was a tall, blonde, athletic life-guard— very sultry, a college senior. In the first few weeks she complained bitterly, but with admirable frankness, "You are not at all what I expected. I was hoping for somebody better-looking. Somebody more important, too. This office is nice, but somebody else's name is on the door. You must be renting space [true], which means, I guess, that you don't have very many patients. I don't think anybody knows who you are. Also, you are only charging fifteen dollars a session, so you couldn't be very good, or in much demand. You're shorter than me, too.

You're not good-looking enough to have fantasies about. What you say is not all that clever, or even terribly interesting. Nothing you have said makes me think that you have any particular gift for understanding me. I don't feel any genuine empathy, even sympathy. I don't think you care that much about me. I don't think I would even look at you or talk to you outside of this setting. Typical for me: I have no luck." Well, this was realistic enough. In fact, the two of us could hardly have been more different. There was nothing terribly distorted in her vision of me.

I wondered how we would proceed. Would I be treated, yet again, to a direct demonstration of the power of transference? I was. Within a month she was "in love" with me. I was everything she had ever wanted in a man. I was handsome, clever, interesting, in great demand, and, most important of all, I "understood" her and obviously cared about her. In fact, she thought, I was probably in love with her as well, though she knew professional ethics prevented me from declaring myself, or acting upon my desire. It was really quite astonishing. I knew that part of my work would consist of getting her back to her original position. If after a year of work, she could feel about me more or less the way she felt when she first walked into my office, then I would have achieved my goal of "resolving the transference." The illusion would cease, she would be "disillusioned" and see me more or less as I was, without the interference of her own needs—which is more or less the way she saw me when she first came to me. She would, perhaps, end up knowing a great deal more about how easy it was to fall prey to these kinds of illusions, in or out of analysis, but that was no guarantee she would not be subject to them again.

At the time, I confess, I did not think along these lines. But I knew for certain that I was no better off in any way that I could see than she was. I was different. My unhappiness took other forms. My ignorance took other forms. My blind spots were other than hers. But in no essential way, in fact, could I claim to be any better than a single patient I had ever seen.

True, I was less unhappy than some. But I was also less happy than others. Many of my patients had a greater zest for life than I did. Could I ask them how they did it? Could I confess this to them? No, clearly. An analysis, even a psychotherapy, could not succeed, I knew, if the situation was democratic and egalitarian. My patients were looking to *me* for answers. They would not welcome a shrug of ignorance and a question in return. And so there was a constant push for me to be wise.

Wisdom, in this technical sense, is not all that difficult to come by. When you have read enough and been to enough case seminars, you know what is expected, what *sounds* profound, what gives comfort, what appears insightful even if it is not. So much of what I said came from my head, not from my heart. But if it had been any different, I would have been exhausted at the end of a day.

This often made me wonder. Here I was, a professional. I had a degree that said so. I was a member of the International Psychoanalytical Association. I was a member of the San Francisco Psychoanalytic Society. I had the right to call myself a "psychoanalyst." I could sit quietly in my chair, at the head of a couch and listen to patients spill their lives into my lap. I could muse out loud. I could make interpretations, and some of these interpretations appeared to be filled with "insight." But I was, when it came to living, and to lives, as ignorant as the next person. Was it just me, I kept asking myself over and over? Were other analysts any wiser? Well, I hope so. Or rather I should say, I hoped so.

As I saw my first patient, I began to discover something far more sinister than the transference. It is called the countertransference. All it means is that the analyst has just about as many feelings for the patient as the patient has for the analyst, only the analyst is not obliged to give voice to them, either to the patient or to himself. For example, I didn't particularly like this patient. She was attractive, and I could have had sexual feelings for her, no doubt, but she was not the kind of person I would have wanted to spend any time with. I was getting worried about this. And so I consulted a senior analyst for a supervisory session. I explained the problem. He told me this was the countertransference. "But why?" I asked. "Why isn't it that I just don't like her?" He explained to me that in the sacred space of the analytic room, "The analyst does not have normal feelings. Everything that happens in that space is to be analyzed. Look, Jeff, it's not just that you don't like her, the way you would not like somebody on the street. You are an analyst. You cannot afford such simple feelings. Besides, if that were all there were to it, you would not feel yourself sexually attracted to her." Now it was my turn to set him straight. "Wrong, that is part of my problem. An old problem. I'm all too often attracted to women with whom I have nothing in common. This is just more of the same."

"Well, then, Jeff, you are incompletely analyzed. You have never overcome this problem, and that is why it is surfacing now."

"Fair enough, but let's at least not blame her. By calling it my countertransference, it makes it sound as if she did something to create this problem in me. She did nothing. Except she's there. It's my problem, one I've always had, and maybe always will."

"Well, maybe you shouldn't analyze anybody then."

"Yes, I've been thinking the same. But tell me, don't you have problems of your own, that also surface in your sessions with patients, or did you analyze them away?"

"Of course I do. That is why I have followed Freud's suggestion, which most analysts take in jest, that one should be reanalyzed every five years."

"Analysis is your whole life, then?"

"That's right."

I couldn't help but feel there was something a little bit odd about a practice that seemed to feed off itself in this way. True, maybe I just wasn't analytic material, either as a patient or as an analyst, but I was beginning to wonder if that was such a bad thing after all.

Chapter Eight

ANNA FREUD AND I

1980

In psychoanalytic research, there was one great historical prize, the unpublished letters between Sigmund Freud and his best, perhaps his only, friend, Wilhelm Fliess. Everybody wanted to lay hands on this treasure. Partly this was because of the simple intrigue: what was inside this document, so assiduously guarded and kept hidden from public view for so many years? My interest was not just the result of my particular area of concern; every analyst was interested in these letters. They were the basic document as it were, the wellspring of psychoanalysis.

When Anna Freud published an expurgated version of the letters in 1950, she simply whetted the public's appetite, and caused rumors to fly wildly. Why had she omitted 158 letters? What was in them? While I was desperately curious to know what they contained, and I had had some encouragement from Eissler, I also realized that I was in no position to even dream of seeing them. Apart from my being so junior in the hierarchy, I also knew not a word of German. What would I do with the letters if I had them, when I could not understand the language they were written in? Knowing German was the key to doing any kind of serious historical research in the history of psychoanalysis. There was nothing one could do without it. If I was serious about my ambition to do just such research, then the very first step was to learn German.

As luck would have it, I met a woman who had been born in Berlin and was now living in Berkeley, and she was interested in an exchange: if I would read psychoanalytic texts with her and talk about them, she would read Freud with me in German, and I would learn the language in the process. Marianne Loring and I began working, and I soon realized that learning German was not something I could undertake casually. It could not be done without a major commitment. And yet I wasn't sure I could make that commitment at that moment. Terri and I had been having serious difficulties in our marriage. The complexity of it was such that I could not fathom how analysis could ever make sense of these kinds of situations. To this day, ten years later, I still don't know what exactly happened, and who was responsible. I spoke to a number of analysts. Each one thought he had understood the situation exactly. But it was simply not possible for each of them to be right. And their "interpretations" struck me as hollow. My friends were no more helpful, but certainly no less.

I had a sabbatical coming up from the University of Toronto, one free year, where I could do anything I wanted. My idea had been to remain in Berkeley and build up a practice. But that was clearly not what I wanted. In fact, what I wanted was to spend the year in Germany, learning the language and poking around archives. I told that to Terri. "Well, why don't you just do it, then?" she said—such liberating words.

And so that summer, Terri, Simone, and I set off for Europe. Our first stop was Cannes, to visit my parents, who lived on the coast in the South of France. Terri and I spent a few weeks together in Corsica, and then I bought a bicycle and set off for Germany, while Terri and Simone returned to Berkeley.

I rode along the French and Italian coast, over the coastal alps in Genova, around the Lago di Garda, and over the Brenner pass into Innsbruck and then to Munich.

It was a strange feeling to be in a foreign city, completely alone, knowing nobody, and not speaking the language. I was determined, right from the beginning, that I would learn German as quickly as possible. By the next day, I had contacted some of the analysts there, and found a charming small apartment on Prinzregentenstrasse that soon filled up with books. I was building a library of German books that would serve me well in the years to come. I went to the Goethe Institute, saw a lot of German movies, talked to a lot of people daily, and slowly I began to see that I was making progress.

Languages were easy for me. Besides Sanskrit, I had spent two years learning Pali, the language of early Buddhism (used in Sri Lanka, Thailand, Burma, and Cambodia). I also studied Tibetan at Harvard, though here I was less successful. It is a non-Indo-European language, and I possessed little talent for mastering it. Today I can no longer read even the alphabet. I had learned fluent French when I lived in Switzerland and Paris, and also Spanish, when we lived in Montevideo, Uruguay. I had briefly studied some Italian, some Portuguese, some Bengali, and some Hebrew. So learning German became an absorbing pleasure for me.

The focus for my language learning was certainly psychoanalysis. It would have been impossible for me to do serious research work in psychoanalysis without learning the language that Freud wrote in. Too much material existed in German that was untranslated or even unpublished. I wanted to meet German, Swiss, and Austrian analysts, partly out of curiosity, but also because I knew that I would need their help later in doing the kind of research I was interested in. I would need people to send me articles, and I would have to turn to native speakers of German for nuances in meaning, even in Freud. James Strachey's elegant translation, known as the *Standard Edition*, is an extraordinary achievement. But Strachey was not entirely fluent in German, and, as I later learned, his work contains a great many simple errors. There are many passages where Strachey simply did not understand the German. For the kind of work I hoped to do, it was essential that I be able to detect these errors.

Soon after I arrived, I talked to a number of German analysts about my interest in the Hungarian psychoanalyst Sandor Ferenczi, and the Munich Psychoanalytic Society asked me to address them on the subject of Ferenczi's last days. Ferenczi was Freud's favorite disciple and a man of exceptional clinical skills and great human warmth. I was, like many analysts, drawn to him. The more I looked into his life and career, the more reasons I found to admire him. Toward the end of his life he had come to believe that Freud was wrong to discount his patients' accounts of sexual assault in childhood. He was hearing these same accounts from his patients, but even more important, from a purely evidentiary point of view, they were confirmed by other patients who confessed to having perpetrated such assaults. Ferenczi was considered paranoid for believing his women patients; the men's confessions were not even discussed. Ernest Jones, the powerful English analyst who had been Ferenczi's analysand, now took up the cudgel against him in deadly seriousness. Jones let it be known after Ferenczi's death in 1933 (he died a few months after the quarrel with Freud) that he was really a homicidal maniac. While I was in London working in the Jones archives I discovered what this really meant: Jones believed that to disagree with Freud (the father) was tantamount to patricide (father murder). And so, because Ferenczi believed that children were sexually abused and Freud did not, Ferenczi was branded by Jones as a homicidal maniac, and this piece of scurrilous interpretation stuck. The burden of my paper was to show how these kinds of rumors, these pieces of malicious gossip, became encoded as scientific gospel by credulous and intellectually lazy analysts. It was not a bad paper.

I had made friends with a child psychiatrist who was head of a university clinic. He was urbane, a gourmet and a bon vivant. He was also in analytic training. I went skiing with his family in the Italian alps, and we got on fine as long as we did not talk about psychiatry. By the time I had finished addressing the Munich Psychoanalytic Society, we were no longer on speaking terms. He got up at the end of my paper and said in a voice filled with emotion, "Your paper shows that you are as paranoid as Ferenczi. I am, as you know, a child psychiatrist and I know that children do, in fact, invent tales of sexual abuse.

Freud was right. Ferenczi was wrong. So Jones was correct too, even if for the wrong reasons. The fact that you, Jeff, can take up Ferenczi's views after all these years of clinical wisdom has demonstrated there was no truth in them shows—well, I don't know how to say this, but I feel forced to say that you are dangerously mentally ill. In fact, Jeff, I believe you should spend some time in a psychiatric hospital. I have to go even further: I am prepared to commit you tonight if one of the gentlemen in the room will second my opinion." I laughed appreciatively. It was a good performance. Quite convincing. He made a point far more dramatically than I could have with my sober words. When I finished laughing I noticed he was not smiling. "Jeff, I'm serious."

So here it was. Fifty years after Ferenczi, to go against conventional "wisdom" was still dangerous. I was being told that a man could get himself locked up that way. The few people who remained loyal to Ferenczi maintain that he died of a broken heart, abandoned by Freud and all the people who had apparently so loved him. I was more fortunate: none of my critic's colleagues was willing to sign a committal form. Displaying a strain of endemic naïveté, I decided that my friend's problems were again intellectual provincialism. He was peculiar, he was different, he had completely misunderstood psychoanalysis. He was not a product of the system, he was a mistake. So I merely had to avoid such people. He was the proverbial bad apple in the otherwise healthy barrel. I could not tolerate the idea that it might be the barrel that was rotten. Of course I could see that the world was full of "bad apples," but I had to believe that the barrel was fine, otherwise my entire training and everything I had believed until now could be in question. All that was needed was to move beyond where I was, literally. It was time, then, to expand my circle.

I wrote to Kurt Eissler frequently from Germany and told him the details of my progress in research. He was pleased, and proposed that I return to London to visit Anna Freud and let her know about my progress. This was not my first meeting with Anna Freud. In fact, I had first met her in London in 1973 when I was still a candidate in psychoanalytic training in Toronto. I was with Terri. Both of us, at the time, were absorbed with issues around trauma, and especially the trauma of the Second World War. We were both deeply interested in the history of World War II and the fate of the Jews. I was puzzled that so little had been written about the war from a psychoanalytic perspective, because at that time I still saw psychoanalysis as concerned with uncovering the effects of trauma.

When we arrived at the lovely old house in Maresfield Gardens (I had no idea at the time how large this house would loom in my subsequent life), we were met at the door by the housekeeper, Paula Fichtl, who was, like Anna Freud, in her seventies and had been with the Freud family since she was a young woman in Vienna. At the time of our visit, Anna Freud was still seeing patients, and we were taken upstairs to her office. She was dressed, as she almost always was, in a loose-fitting, long dress and wore many necklaces. As soon as we sat down she said, "Well, what do you want to talk about?" Clearly no small talk was going to be tolerated here.

"Terri is Jewish, she survived the Warsaw Ghetto, and I am also Jewish, and both of us have been immersed in holocaust literature. We are puzzled why so little has been written about the holocaust in psychoanalysis."

"I am puzzled by your puzzlement," she said immediately. "Why should psychoanalysts in particular write about the war?"

"Because so many Jewish analysts are refugees from Nazism."

"But that has nothing to do with psychoanalysis."

"But doesn't trauma play a central role in analytic theory?" Anna Freud shrugged her shoulders, apparently dismissing my concerns as uninteresting. I was deeply disappointed.

Anna Freud was Jewish and had come from what I imagined to be a cosmopolitan environment (it had simply never occurred to me that Freud could himself have been provincial—

indeed it shocks me to say it even now). I had always assumed that the lack of interest I found among my psychoanalytic colleagues in Toronto was purely a product of the provincial nature of Canadian intellectual life. It was not, I thought, connected in any way to psychoanalysis. Thus I fully expected Anna Freud to be as enthusiastic about this topic as I was. That the person bearing the greatest living name in psychoanalysis should be so uninterested was disheartening. I felt we had come in vain.

Anna Freud folded her hands and sat expectantly. I thought the interview was over, since clearly Anna Freud did not much take to us, and I felt deflated by her apparent lack of interest in the Second World War.

I tried again.

"I know that your father never wrote anything about the Nazis, but he must have talked to you about it. What did he say?"

She simply shrugged her shoulders, and sat silently. I could not tell if she meant that he had told her nothing, or if she did not intend to tell me anything. I tried again.

"I don't mean to pry, but I know from Jones that you were questioned by the Gestapo in 1938, and that you do not want to talk about what happened that day. Is this why you would rather not speak about the holocaust and psychoanalysis?"

"Not at all. I just don't think that psychoanalysts have any more reason to be interested in the holocaust than anybody else. How, for example, has it affected psychoanalytic theory?"

"Well, in some ways it has affected it negatively. I mean it occurred to me that somebody like Melanie Klein may write the way she does simply in order to avoid facing issues derived from her own past."

Anna Freud suddenly perked up. I had piqued her interest. But not for the reasons I thought.

"What do you mean, exactly?"

"Well, her case histories are so denuded of personal detail that you cannot even determine where she conducted an analysis, in what language, and when, as if all these historical details were simply too trivial and extraneous to the core of analysis to be worth mentioning."

I was unaware at that time what a bitter rivalry had separated the two grand women of psychoanalysis in London—in fact loyalty to one or the other determined what institute one joined in England and thereafter one's entire subsequent training and career (though some opted to join a neutral group).

Anna Freud seized on my comments, ignoring the theoretical point I was trying to make, and launched what was for her an exceptionally spirited attack on Melanie Klein, something she never did in print or in public.

"I won't repeat the gossip about her analysis of her own children, because such personal attacks have no theoretical value. Many of her written cases come from her own children. But you should read Edward Glover's article about the differences in our approach. You know that he was first a Kleinian, but after analyzing Melitta Schmideberg, Klein's daughter, he changed his mind? My father, too, you know, did not think that Melanie Klein's ideas were compatible with classical psychoanalysis."

We could agree on this, although I suspect now that it was for very different reasons, and I felt emboldened to say something about myself.

"I cannot tell you how much I admire your father's writings. I am in analytic training in Toronto, and it pains me to see how little the analysts there care for his work. Terri and I were just in Paris, and it was such a pleasure to meet analysts who do respect your father and seemed so much closer to him in spirit than the Toronto group. I was so pleased to hear the presidential speech by Leo Rangell [the president of the International Psychoanalytical Association for several terms running] using psychoanalysis to examine the compromise of integrity. And meeting Edward Weinshel and Victor Calef made it feel like the good old days, when important battles were still being fought around Freud." I didn't realize that I had touched a hornet's nest.

"You are completely wrong," she countered. "The good old days are gone. And the battles that many of the senior people in the International are busy fighting have to do with attempts to prevent my child study center from incorporating as an institute in the International Psychoanalytical Association. Well, if you have no further questions . . ." and she rose from her chair, extended her hand to both of us in farewell. I fully expected never to see her again.

It was now many years later. I was no longer a candidate, but a psychoanalyst, and I came highly recommended by her great friend, Kurt Eissler. Anna Freud was decidedly more friendly, yet cool and distant even so. She still wanted to get down to the business at hand, and not waste time on trivialities. For a psychoanalyst, she was remarkably closed on many issues that one would expect her to be open to. Many people had warned me against asking personal questions. She did not welcome them. But as I began to spend more time with her, I also had the strong feeling I was being remiss in my historical obligation if I did not ask her certain questions about her father. Of whom else could one ask these questions? And how many more opportunities would I have?

One day Anna Freud and I were going through Freud's desk, and I found a large envelope that contained passports, photos, and other documents from Freud's earliest childhood that she had evidently not seen for some time. She took a keen interest in going through them with me and seemed so happy to answer my questions about minor matters that I decided to ask something I had always been curious about.

"Miss Freud, your father wrote, in one of his early papers, that he thought it a pity that children were rarely enlightened about sexual matters. I take it then, that he talked to you and his other children about masturbation, is that correct?"

She stared at me aghast and in silence. I tried to redeem myself. "Your father did write that such matters should be discussed."

Shaking her head, she replied. "That was not for the family."

Paula, who often came into the library when Anna Freud left me there to work, was a far more engaging conversationalist, especially about Anna Freud, whom she seemed to dislike intensely (their dislike was mutual). She spoke to me in a mixture of Austrian German and broken English. "Things have never been the same since the professor died. He was a good man. A very good man. He liked me. He teased me. He told me I was a better judge of character than most psychiatrists. I always told him who of his patients I liked and who I didn't. He told me I was always right. He was fun. Not *her* though. She's no fun. And she's very cheap. Don't let her see me giving you all these cakes. She tells me not to. She wants to eat the same thing all the time, at the same time, too. And no eggs in anything. How can you cook without eggs? I slip them in anyway, but she sometimes guesses and won't eat it. Have you seen her room? I'll take you up there later. It's like a little girl's room. Animals—how you say, *Plüschtiere* [stuffed animals] everywhere. You know, Fräulein Anna has never once touched me— not even my hand, like this, see? Maybe his fault though. I never saw him hug or kiss any of the children."

Anna Freud certainly gave off an aura of physical coldness. Rumors abounded concerning her sexual life. Nearly everybody close to Freud (e.g., Ferenczi) or not so close (e.g., Siegfried Bernfeld) was said to have been considered a matrimonial match for her. In fact she never married and spent the last years of her life with Dorothy Burlingham. Nobody knew if they were lovers. Anna Freud struck me, and many other people as well, as completely asexual, and I would not be surprised to learn she and Dorothy Burlingham were lovers without engaging in any sexual intimacy. I asked Paula about it once, and she said they did, sometimes, share a room. She smiled as if she knew something, but I doubt that even she knew.

Paula's greatest bitterness was reserved for a struggle over a magnificent coin collection that Freud gave her. "After he died,

Miss Freud told me it wasn't really a present, and that I had to give it back. But I won't, ever. The professor gave it to me, and I am going to keep it."

As I spent more time with Anna Freud, we became more familiar with each other. She was more willing to talk with me and tell me her views after it became clear that Eissler wanted me to eventually become his successor as director (his official title was Secretary) of the Sigmund Freud Archives. He had first broached the topic in a letter to me while I was studying in Munich in 1980. When I returned from Munich, he met me at the airport in New York and we went to his house, where he told me in great detail about his plans for the future of the Archives, and how he wanted me to take over. He thought it a good idea for me to return to London as often as possible and to work as much as possible with Anna Freud in Freud's house.

Eissler liked to tell people that I was an "idealist." By this I think he meant that I was never all that interested in merely making a comfortable living pursuing a psychoanalytic practice. This was no minor incentive for many analysts. After all, it is not now unusual to charge $150 for a single fifty-minute session, and most analysts see at least five or six patients a day—many up to ten per day—so that over the year, an analyst in demand makes a handsome salary. But perhaps because of my academic background, I was drawn to historical research instead. This, in turn, recommended me to people like Masud Khan in London, who was one of the directors of the Sigmund Freud Copyright, which owns the copyright on all of Freud's works, and to Mark Patterson, a literary agent and the person responsible for the day-to-day workings of the Copyright. The Freud Copyright gives all permissions, worldwide, for the publication of any work by Freud, or for quotations from his works. It also has a major say in deciding what unpublished material will be handed over to somebody for publication. I was asked to join the Copyright as one of the directors. This was in 1980, after Eissler had asked me to join the Archives.

Masud Khan had already heard that Eissler wanted me to

succeed him in the Archives. He felt, correctly, so I believe, that it would be a good thing if the Sigmund Freud Copyright and the Freud Archives could coordinate their activities. Apparently Eissler had never been entirely cooperative, and Khan was curious to meet me. While I was visiting Anna Freud in London, he invited me to tea in his sumptuous London apartment. It was the fist time we met. I remember that while we talked a beautiful and voluptuous woman wandered through the room without saying anything, dressed only in a sheer nightgown. Khan explained that she was a French actress and was staying with him. He was eccentric, and seemed impressed with my having been a professor of Sanskrit. He loved everything French, and it appealed to him that I was part French and spoke the language. We seemed to get on well at first, but it was in fact a mutual illusion. The more we learned about each other, the less we liked each other.

Once he said to me, "You must be Jewish, am I right?" "Yes, I am, why?" "Hmmm, it figures." I thought he was paying me a compliment, but later I heard, from one of his patients, that he was viciously (though eccentrically) anti-Semitic, and this was borne out by his book, *The Long Wait*, published just before he died in 1989. I quote one particularly offensive passage:

> I warned Mr Luis: "One more personal remark about me, my wife, my staff or my things, and I will throw you out, you accursed nobody Jew. Find your own people then. Shoals of them drift around, just like you. Yes, I am anti-Semitic. You know why, Mr Luis? Because I am an Aryan and had thought all of you Jews had perished when Jesus, from sheer dismay—and he was one of you— had flown up to Heaven, leaving you in the scorching care of Hitler, Himmler and the crematoriums. Don't fret, Mr Luis; like the rest of your species, you will survive and continue to harass others, and lament, and bewail yourselves. Remarkable how Yiddish/Jewish you are."

Khan was regal in more than his bearing. He was the prince of a small kingdom that was now part of Pakistan and had inherited a vast fortune. He had a staff of servants, and he commanded them as if he were still living in the middle ages. On the other hand I met patients of his who told me that he was unbelievably kind to them, especially when they were in distress. I am not sure what they meant. For one woman, at least, it meant taking her into his home. It was commonly rumored that Khan slept with his patients. At one point in our first conversation over tea he asked me to tell him what gossip I had heard about him. I told him, "Well, for one thing, I was told that you were haughty and arrogant and thought of yourself as above most people."

"Well, that's certainly true enough. What else did you hear?"

"I heard that you slept with patients."

"Yes, well, let me tell you what I told [X, a well-known American analyst] when he was here last. I told him, 'The story about you is that you sleep with other people's patients.' 'You're a fine one to talk, Khan,' he shot back. 'Ah, but that's different,' I told him. 'I never sleep with other analysts' patients, only my own.' " And here Masud Khan howled with laughter in what seemed like good-natured self-criticism. Only later, did I reflect that the statement contained not a shred of self-criticism; moreover this was not just an amusing story. Women had come to him in distress and trusted him, while he saw it as the stuff of anecdotes to misuse that trust and sleep with them. I heard, later, that not only was he living with a former patient, he continued to see her husband in analysis.

After Anna Freud and I became more friendly, what seemed to animate her most was discussion of the sorry state of affairs into which psychoanalysis had fallen. She had nostalgia for the good old days, when Ferenczi, of whom she seemed particularly fond, was around. She spoke sadly of a decline. It amused her to hear me berate the many bad books written in the field (little did I know then that she would have regarded my own book, The Assault on Truth, as far worse than anything yet written). A subject that concerned her a great deal was her edition of the Freud/Fliess letters. She would insist that nothing of significance had been omitted, and when I tried to argue she would become upset. Freud's patient Emma Eckstein loomed large in my conversation, because I knew that the whole story of her analysis with Freud and the disastrous operation with Wilhelm Fliess had not been told, that there was still a great deal to learn. But all she would tell me, mysteriously, was that Emma Eckstein's name was *"ein Begriff* (familiar to her)"; when I asked her to elaborate she would fall silent. My frustration was often great, because I yearned for explanations of historical matters that only she knew about and I knew that people were intimidated by her and would usually refrain from asking her questions, especially on these matters. "Did your father speak to you about Fliess?" She shook her head. "Do you think Fliess's letters to him might still exist?" She shrugged. "Do you think your father still thought about Fliess in later years?" Silence. Anna Freud did not encourage curiosity. She herself, I felt, was singularly devoid of historical curiosity, although she would happily talk for hours about case histories and contemporary analytic theory. We were clearly mismatched.

If there was one area that she was more private about than any other, it was her relationship with her father. There can be no doubt of its intensity. Paula once told me that when Freud was too sick to be moved from his study, Anna Freud took up residence there, displacing her mother, Martha. She cared for him with all the intensity of the daughters Freud describes in his *Studies on Hysteria*. Often when I was alone in the house I speculated about what treasures might still be hidden away. Would it be possible, I wondered, to find Fliess's letters to Freud, which Freud said he had hidden away so ingeniously that even he could not locate them? Would the "big dream" of the *Interpretation of Dreams*, the one Fliess asked him to remove for the sake of discretion, and which is presumed lost, turn up? What about the truly personal letters, such as those between Freud and his favorite child, Anna?

So many rumors abounded about Freud and Anna. It was

known that Freud had analyzed his own daughter, because he wrote a letter to the Italian analyst, Edoardo Weiss, warning against attempting the same with his son, even though he wrote him that with Anna "it had turned out well"; and that her hand had been sought by many analysts close to Freud, even Ernest Jones, but she never married. Moreover, her personality was so severe that few dared to ask her personal questions. Had she ever loved a man? Did she love women? Specifically, did she have an affair with Lou Andreas Salomé, a lover of both Nietzsche and Rilke, and later a disciple of Freud? Did she live with Dorothy Burlingham (of the Tiffany family, and a woman who was analyzed by Freud in 1925) because she was in love with her? What did she think of her analysis with her father? These and many other questions could never be answered by asking Anna Freud. The hope was that Freud's letters to her and her letters to him would answer such questions. The letters, I later discovered, are singularly devoid of personal material and bland; they are astonishing for a reason I had not counted on: they are completely vacuous. In the letters to Fliess, and even in the letters to Jung, Freud is engaged and personal, whereas the founder of psychoanalysis seems to have had nothing of importance to say to his own daughter. I really don't know why.

One day while I was still in Germany, Anna Freud invited me to a conference at her home in Hampstead. Many of the analysts from around the world that I had long admired were going to be there, and I was eager to attend. There was something strange about meeting in the actual house where Freud had spent his last year. The large, imposing brick house in Hampstead was no different, really, from many of the other houses on the block. But once inside, there was no mistaking that it was a house filled with memories of the past, of better days. I remember that the first time I went into that house I was struck by a certain smell, a palpable mustiness. It was very dark and almost completely silent. As soon as I entered, I felt weighted down. Mementos were everywhere.

The quiet somber quality of the Freud house did not fit in with the meeting itself. Anna Freud herself shared that quality, as did many of her elderly women friends who were there: Jeanne Lampl de Groot, Muriel Gardiner, Anny Katan. Eissler was very solicitous of all these women. One day he reprimanded me for not helping one of them cross the street. "Professor Masson, how can you be so callous? Can you not see the lady is having trouble crossing the street? Run, and help her." He was right, and I did. But most of the analysts there were from America, and many, especially the males, were in one sense or another "on the make." So there was a certain frenzy in the atmosphere. Alliances were being forged, enmities hardened, dominance expressed. Everybody was carrying around large black notebooks and noting down the dates of the next conferences, trying to get invited to the "important" ones and avoid being seen at the others. It was autumn, and there was a certain melancholy in the air around Hampstead at that time of year, but it was lost on these prestige seekers from another continent. Had Eissler not been there, I would have felt very lonely.

An American professor of child psychiatry gave a paper about a particularly lovely woman patient whose analyst fell in love with her—but this is not how the professor told it. He told it in analytic terms, so that what he discussed was the patient's attempts to seduce the analyst, and the countertransference difficulties this caused the analyst. He kept avoiding talking about any real feelings the analyst had for this entrancing woman. Yet it was obvious that the man had simply fallen in love. (Of course I couldn't help wondering if the analyst had not been the professor himself, since he seemed to know so much about the matter.) As I walked out of the room I turned to Dick Newman, an older analyst from New Haven. "Dick, didn't you get the feeling that this analyst had simply fallen in love with his patient?"

"No question about it, Jeff, it was obvious to everybody."

"But why didn't anybody say anything then?" Dick just shrugged, as if I was missing the obvious. Truth, clearly, was not the main point of this conference; political alliances were far more important.

When it came time to give my paper, I talked about the new material relating to the sexual abuse of children in the letters of Freud to Wilhelm Fliess that I was editing. The American professor began the discussion by saying, "This is a very provocative talk, and I feel provoked." Leonard Shengold came to my assistance, but from a certain distance. When I thanked him for it later, he told me I should not be grateful, because he had to force himself to do it. "Jeff, I have to tell you that I agree with what you say, but I deplore your way of saying it." Shengold was a heavy-set, older analyst with a friendly demeanor who in spite of his size looked younger than his years. We had been friends for years in the way that analysts are friendly with one another, from a distance and somewhat remotely. I respected him, and I knew he was capable of rising above all the political back scratching of this particular conference. So although I felt personally abandoned by him, I had to reflect on what he said. Was it possible that the analysts could not hear my "message" because the messenger was so obnoxious?

At the plenary session I made some comments about feeling honored at being there, and wished to return the compliment by reading to the assembled guests some unpublished lines I had found by Freud. From the audience came the words, "Never mind the words, just give cash." Despite the vulgarity of the remark, he had a certain point. There was a kind of pretentiousness to my comment. I was by far the youngest analyst at the conference. I was only there because Kurt Eissler wanted me there and perhaps Muriel Gardiner and Lottie and Dick Newman. Why was I pushing myself forward? Ostensibly because I had found some new information about Freud. But did I have to be so eager to display it to everybody who didn't ask?

The conference lasted for four days. One evening I approached Muriel Gardiner, to whom Eissler had just introduced me. It was the first time we met. I knew that she was a good friend of Kurt Eissler, and of Anna Freud. I had admired her work on the Wolf Man. She was, apparently, the model of Lillian Hellman's heroine Julia, in *Pentimento* (later made into the film *Julia*), and a woman of enormous courage. She had been a medical student in Vienna in the thirties, and a member of the Freud circle who had used her immense inherited wealth to help members of the socialist underground and the Austrian resistance hide and escape, taking considerable personal risks. (She explains this modestly in her book, *Code Name Mary*). She became a psychoanalyst, but a somewhat unorthodox one. She was universally liked and respected.

I told her my feelings about the meeting. She responded in kind. She was very direct and personal. I remember her words. "Jeff, you strike me as a good researcher. I think you are on to something, the material you are finding in the Freud house is valuable. But the world of clinical psychiatry and clinical psychoanalysis is just not for you. I can see that you are having a tough time here, where there are few researchers, and many clinicians. Maybe you should think of doing pure historical research and give up the idea of practicing analysis. That seems to be both your talent and your interest, aren't I right? Maybe you just don't want to do therapy." She was right, and my resolve to follow my interests was strengthened by her words.

I now began to see Anna Freud more frequently. At the door I was generally greeted by Paula, who was always happy to see me, because it gave her the chance to talk and gossip and tell me all the news since my last visit—who had been there, what had been going on. I mentioned earlier the distaste that seemed to exist between Paula and Anna Freud. They inhabited a kind of war zone. Paula insisted on speaking to Anna Freud in English, which, many years after arriving in England, Paula still barely knew. The first time I heard Paula say anything to Anna Freud, Anna Freud turned to me and said, loudly, "I can't understand a word she is saying." Truthfully, neither could I. Paula returned the dislike. When Anna Freud was not present, she would come into the room where I was working in the house, bring me tea, and sit down. I told Eissler what Paula had once said to me about Anna Freud never hugging her. He explained that it would be unheard of, in the Austria of Freud's day, for a "master" to embrace a "servant." Anna Freud was a product of that era. But that meant that analysis had never freed her from this false position. (How many of us are ever freed from anything, by analysis or any other method short of falling in love?)

Paula told me that she never had a day off, never went anywhere, never had any fun. That I can believe. The house was not a joyous place. (But then it could not have been much fun for Anna Freud to be an icon either). The only lively creature there was Anna Freud's chow, a young and frisky purple-tongued delight. I often took him into the large garden for a run. Anna Freud loved it when I did that. She would lie on the deckchair and watch us, and afterward we had our best talks together. We both liked dogs, and for some reason this allowed her to open up to me a bit. On this occasion (1980) I reminded her of our first meeting, seven years earlier, and how I had not understood her subsequent warning about the corruption rampant in psychoanalysis. I told her something about my own analysis, and my experience with my supervisors and teachers in psychoanalysis in Toronto. "I am not at all surprised," she said. "Yes, it is really quite terrible." "Has it always been like that?" I wanted to know. "I don't think it was like that in the early years," she answered, "but I know that if my father were alive now, he would not want to be an analyst." Of course we both had our own reasons for believing that corruption was all-pervasive in analytic circles; on the other hand, we were probably both right.

Once Anna Freud trusted me, there was no end to her generosity in letting me take whatever I wanted out of the house for copying or studying. She did not seem the least bit attached to the volumes in her father's personal library, and I certainly could never have done the research I wanted to do without her full and unstinting cooperation. She was also always prepared to answer my many questions, as long as they were not personal. The very first time we talked about the seduction theory, I told her that Freud had once claimed that he gave it up because he read, in Havelock Ellis, that people could be sexually assaulted, and yet escape neurosis. "Could we look for Ellis's book in your father's library, and see what he wrote in the margin?" Anna Freud's eyes shone with pleasure. "Yes, that would be interesting. You know, Dr. Masson, sometimes I do understand your research zeal, even if it does border on the fanatic." We spent the next couple of hours going through Freud's books and reprints by Ellis, but could find nothing about sexual seduction. "Miss Freud, this is really odd. I wonder what your father had in mind?" "Maybe you should look at other books by Ellis, besides the one my father cites. Maybe he made a mistake." I did, and could find nothing about seduction or its effects in all the seven volumes of Ellis's *Studies in the Psychology of Sex.* This was one of the first clues I had that Freud's own account of his abandonment of the seduction hypothesis was not entirely reliable, and that the full version would have to be dug for.

During the summer of 1980, Eissler told me that the Sigmund Freud Archives, with funds from Muriel Gardiner and the Newland Foundation, had recently purchased the Freud home. The idea, he said, was to eventually turn it into a research center (at that early date there were no plans to make it a museum as it now has become). Part of my job, as his successor at the Archives, would be to live in the house and to oversee the conversion. I was to move there upon Anna Freud's death, and to make certain that the hundreds of letters there were catalogued and eventually made their way to the Library of Congress. The house was unchanged since Freud moved into it in 1938, and I made plans to convert it into a brighter place by putting in skylights and to make the magnificent library available to researchers. My responsibilities were to conduct research, but I would be fully independent.

While the general atmosphere in the institute and the house was bleak, I loved, and could not get enough of, the discoveries I was making there. For example, Anna Freud showed me a case history dated December 22, 1897, never published, in which Freud wrote to Wilhelm Fliess about an early childhood sexual assault. It was one of Freud's most direct and detailed descriptions of child sexual abuse, an altogether remarkable case history about a small girl who nearly died because of an assault by her father. Freud ends the letter by proposing to Fliess that from now on the motto of his new science of psychoanalysis should be the moving words from a poem by Goethe: "What have they done to you, poor child?" Nowhere else in the published Freud is there such direct evidence of Freud's exquisite sensitivity to the suffering of children. As I now reread my letters and reports from this time, I can see it clearly: I was intoxicated with the possibility of finding Fliess's letters to Freud, which had been lost since 1904. I was convinced I would find the missing dream from the *Interpretation of Dreams,* the one dream that Freud said he had completely analyzed, but which was removed at Fliess's insistence, because it was too sexually indiscreet. Most important of all, I would find the clues to what really happened with the seduction theory. There was only one catch: nobody wanted me to find out.

Anna Freud listened in stony silence while I painted a marvelous mural of all the hidden truths coming to light; doors being unlocked, things falling into place.

I asked her one day, "Miss Freud, what do you think your father meant when he said that he heard things in the Paris morgue of which medical science preferred not to know? Isn't it a strange reference?"

"I don't know" she conceded.

I was pleased. "I have some news. I think I've found the answer to the puzzle. I was in Paris recently, and found the records for the day your father was present in the morgue. That day Paul Brouardel, the forensic psychiatrist, 'demonstrated' the corpse of a child who had been raped and murdered by her father. So it seems clear that this is what Freud meant. Medical science didn't want to believe this was possible. They refused to admit it. Wouldn't you agree?" Silence. But I was too dense and too caught up in my discoveries to notice her resistance. I would often knock on her door (her bedroom was on the third floor), give another breathless display of my historical quest and dash away to find new pieces to the jigsaw puzzle. The house was like a gigantic treasure chest. Although Anna Freud did not know that I was someday to live in the house, she did know that I would be in charge of research projects for the Archives, and she had given me permission to see anything I needed and to look around for material more or less throughout the house. I do not believe that Anna Freud ever consciously concealed material. She was generous and self-effacing to a fault. But I think she assumed or hoped I had the same priorities she had, in which family loyalty played a central role. There was certainly never a plot on the part of the keepers of the flame to maintain the image of a Freud that was clean, even sanitized, but it was clearly intolerable to many close to Freud to think that he was anything less than morally irreproachable.

It is hard for me to convey the excitement of the discoveries I was making then. I would open a drawer, and Freud's unpublished letters about Ferenczi's apostasy would fall out. Letters between Breuer and Freud, letters to and from Charcot, a few letters from Fliess to Freud, letters to Minna, hundreds of family letters, all of these were to be found somewhere in the house. Anna Freud herself had only the vaguest memory of where things were, or what there was. My joy was so pure, so obvious, that it began to act on her, and she too, got caught up in the search for new documents. There were puzzles everywhere, and not unimportant ones. Why did Freud keep a whole packet of Ferenczi material, all connected with Ferenczi's views about childhood seduction, in the top middle drawer of his desk? Why was it so important to him? Or had somebody else put it there? Who? Anna Freud herself? Dorothy Burlingham? Paula was functionally illiterate, so she could not have collected them. Eissler paid frequent visits to London, but he was so deferential he would never have wandered about looking for things as I was doing.

Anna Freud was remarkably trusting and open. I borrowed hundreds of papers, copied them, and returned them. I even asked her if I could take many of Freud's books with his own marginal pencil notes home to California to study. She did not object, though when I told Eissler about it he was appalled. "What if the plane crashes?" He was right of course, and when I returned the books on my next trip to London I did not borrow any more.

Eissler was not exactly an informal man, but even he found the cold formality of Anna Freud's house something of an ordeal. I remember once we were sitting alone with Anna Freud, and the conversation was bogging down. Anna Freud had a disconcerting way of simply staring at you waiting for you to say something, to which she would often merely shake her head. It was never clear whether this was simply because of her age, or whether she was disagreeing with you, or simply could not believe you could say anything so stupid. Even Eissler was uncomfortable, and when she rose to bring us some more tea, he confessed to me that he wasn't sure what more he could think of to say to her. "Tell her something about your new discoveries about Freud's patients," I said. "She likes anything about her father."

"You're right. That's what I'll do." When Anna Freud returned Eissler began.

"Miss Freud, you will be interested to hear that I have finally tracked down the last of Freud's patients who is still alive. It is the woman Freud described in a case of female homosexuality. She is a very old woman now, but she is delightful. She is interested in everything: she skis, she paints, she plays instruments, she writes poetry, she travels, she keeps up with all the intellectual fashions of the times. She told me she is interested in everything. Except for one thing. Guess what that one thing is."

Anna Freud was silent.

Eissler gave the one-word answer: "Psychoanalysis"— and then burst into what he hoped would be general laughter. I was the only one who joined him. Anna Freud merely shook her head. She was clearly not amused. For Anna Freud there seemed to be absolutely nothing light or amusing about psychoanalysis, nothing she seemed able to distance herself from and laugh at. She was wholly committed to it. She inhabited psychoanalysis the way a hermit inhabits a cave. There was a purity of purpose, a holiness to her devotion that gave off a whiff of religious piety. I did not find it attractive, but it was genuine, and I was impressed. I don't think she invented this trait, either. I am sure she got it from her father, who of course was entirely consumed with holy zeal for the cause. Her father's legacy lay heavy on her shoulders, and she was not about to tolerate any kind of levity. This gave a kind of lugubrious atmosphere to analytic occasions, which contrasted sharply with the life I had imagined for myself. Once again I had to ask myself in all seriousness whether this was a life I could lead.

Meanwhile, back in Munich, I wrote to Eissler to keep him informed of all my progress. I was confident that I was on the trail of a major discovery. And he, too, though he did not like the direction my research was taking, was impressed at the pattern that was emerging. He clearly admired my dedication to scholarship, and also, as he told me, was pleased that I did not represent any faction within psychoanalysis. I was simply curious. This, I like to think, is the reason he wrote me that letter that so stunned and delighted me in April of 1980 asking me if I would be willing to take over the Freud Archives.

Eissler asked me to join him and Lottie Newman for a meeting in London with Anna Freud. We would ask her to release the unpublished Freud/Fliess letters so that I could edit them for publication, and Lottie and I would translate them.

When we met that summer in London, Eissler was at the height of his admiration for the research projects I envisaged doing. We spent hours discussing all the new possibilities for research: how to approach Dora's son, who (understandably enough) wanted nothing to do with psychoanalysts; Freud's letters to Ludwig Binswanger, the founder of existential psychiatry in Switzerland (I was to visit the Bellevue Clinic in Kreuzlingen on Lake Constance in Switzerland); how to get the letters from Eugen Bleuler's son that were still in Zurich (Bleuler had been Jung's teacher, and head of the Burghölzli sanitorium in Zurich); the Wolf Man tapes; what material was still missing from the Archives; should we eventually publish the letters between Freud and his wife, Martha; who was trustworthy, who was genuine, whom we hated, whom we loved. Eissler was apprehensive about approaching Anna Freud. He explained to me that many attempts had been made in the past to get her to part with these most precious of all letters, and that she had always consistently and coldly refused. He wanted to know what my strategy was. I told him I believed that all the material I had been showing her had convinced her that the time was ripe for a new scholarly edition. I was persuaded that my enthusiasm would carry the day. He was skeptical, but hopeful. He wished me good luck.

We all met in her home one Sunday summer morning, in Freud's library. As usual, the atmosphere was stiff at first. There was no pleasant chitchat, just the four of us sitting in a circle on little hard wooden chairs. Anna Freud opened the conversation by saying that she was, in principle, opposed to publishing the letters. She said she could not understand why people wanted these letters so badly. However, she conceded that she had been impressed with the amount of material I had found, and the obvious sincerity of my search for historical truth, even though by then she knew that she did not necessarily agree with the direction in which my search was taking me. She certainly didn't agree with my conclusions. She wanted to know what Eissler thought.

He was brief, but supported my request. He made the logical point that if, as Miss Freud maintained, there was nothing of historic interest in the remaining letters, then the public

could see that for themselves and stop believing that the Freudian cupboard was still stocked with dark secrets. She asked
Lottie Newman if she would supervise the translation, and work
closely with me. Lottie said she would. Anna Freud sighed and
said, "I guess, then, that I have to agree to your editing them.
But I am not entirely happy with the decision."

Now, years later, I wonder if she didn't intuitively understand how sensitive the issue of child abuse was, and how
much light these letters shed on that issue, not always to the
credit of her father, herself, or indeed the profession of psychoanalysis. Anna Freud was no fool, and could plainly see the
direction my research was taking, no matter how reluctantly.
She knew that her love for the man who had been her father
as well as the founder of psychoanalysis might well conflict with
the search for historical truth. Yet to her credit and to my eternal
gratitude, she did nothing to impede that search.

In 1895 and 1896 Freud was not yet a figure of any authority,
and he had no reason to conform to the majority opinion. He
could listen to his patients. This was by no means the norm.
What he heard, and believed, had revolutionary potential. Here
was a man, possibly the first in recorded history, who heard
about the sexual abuse of children and recognized what it really
meant. Women who came to his analytic office told him, with
his encouragement, about the sexual abuse they had suffered
as children. Freud was struck—later he was to say he felt he
had discovered the *caput Nili* of human suffering, the source of
the Nile. This was, indeed, and Freud knew it then, a truth
about what men did to girls, ignored or denied or lied about
since history began. What happens in the minds and bodies of
the girls to whom it is done? What happens to their psyches?
What are the consequences of such immense, silent suffering?
What happens when they cannot tell, because they would not
be believed, or even if they were, they would be censored or

censured? What does this do to some essential trust in the very solidity of the world? Such women live alone in the world. They are alone because they know something they cannot share, and they cannot share it through no fault of their own. Since it is not possible to survive in a universe that is heartless, eventually the denial on the part of so many people in the external world must be internalized as doubt. It is easier to believe oneself crazy, or paranoid, or delusional, or bad, than to believe that the whole world could turn a blind eye to one's real history. It is intolerable to exist in such a world, far easier to accept the blame. It is a situation calculated to drive anybody insane. It is possible to survive as the lone nut in a sane world, but not as the lone seer in an insane world. And yet Freud believed these women—not only in the privacy of his office; he said so publicly in his 1896 paper "The Aetiology of Hysteria":

> All the strange conditions under which the incongruous pair continue their love relations—on the one hand the adult, who cannot escape his share in the mutual dependence necessarily entailed by a sexual relationship, and who is at the same time armed with complete authority and the right to punish, and can exchange the one role for the other to the uninhibited satisfaction of his whims, and on the other hand the child, who in his helplessness is at the mercy of this arbitrary use of power, who is prematurely aroused to every kind of sensibility and exposed to every sort of disappointment, and whose exercise of the sexual performances assigned to him is often interrupted by his imperfect control of his natural needs—all these grotesque and yet tragic disparities distinctly mark the later development of the individual and of his neurosis, with countless permanent effects which deserve to be traced in the greatest detail.

It was a moment of great personal courage. It was not to last.

Freud's courage was not rewarded by his colleagues. While working at Anna Freud's house, I found an unpublished letter in which he told Fliess, less than two weeks after he gave the paper, "I am as isolated as you could wish me to be: the word has been given out to abandon me, and a void is forming around me." Both the immediate response to the paper, and

the subsequent response were ones that Freud had not anticipated: his colleagues thought he was crazy to believe his women
patients. This was bound to have had a disastrous impact on a
young physician with a growing family, eager to open a neu-
rological/psychiatric clinical practice. Where were his referrals
to come from, if his colleagues thought he was completely daft?
I made this point to Anna Freud. "Do you believe," I said, "that
this could have had anything to do with his later abandonment
of the theory?"

"No." She was adamant.

"But tell me, Miss Freud, why did you omit this passage
from your published edition of the letters?"

"Because it makes my father sound so paranoid," was her
response.

"But if it was the truth, then he was *not* paranoid, he was
simply perceptive."

I was slowly learning that Anna Freud could not get caught
up in the excitement of this historic moment to the extent that
I did. It was personally deeply charged for her. Of course this
was her father and his reputation that were at stake. The matter
could hardly have been of insignificance for her, and she could
not be expected to feel disinterest. Nevertheless, the enterprise
of looking back, via the documents she and I were now amass-
ing, to Freud's decision that his patients were telling the truth,
was not without its exhilaration. There exists, as far as I know
(I looked without success), not a single published account of
the devastating effects of incest or childhood sexual abuse before
Freud's time. And yet if this was happening to anything like
the extent that is true today—and why should it be any different?—then at least one in three women, possibly more, in the
general population had been exposed to a forced and unwanted
sexual advance during childhood. In other words, sexual abuse
of one form or another was the core trauma of many women's
lives, yet there was total silence about it. There was no taboo
on the commission of incest, only a taboo on speaking about

incest. For Freud to have broken that taboo of silence was, to my mind, one of the great moments of history.

Later, in one of the most famous retractions in the history of ideas, Freud was to recant. As he put it in 1925 in *An Autobiographical Study*: "I was at last obliged to recognize that these scenes of seduction had never taken place, and that they were only fantasies which my patients had made up." But he had recognized his "error" long before that, in perhaps the single most famous letter Freud ever wrote, one to Wilhelm Fliess, dated September 21, 1897: "I want to confide in you immediately the great secret of something that in the past few months has gradually dawned on me. I no longer believe in my *neurotica* [theory of the neuroses, i.e., seduction]." It was this letter, one that Anna Freud herself published in 1950, to which she continually drew my attention. "Is that not plain enough for you, Dr. Masson?"

It might have been, until I found a letter from Freud to Fliess that Anna Freud had *not* published. It was dated December 12, 1897, almost three months *after* Freud supposedly abandoned his theory that his women patients were telling him the truth. In this letter he writes:

> My confidence in the father-etiology has risen greatly. Eckstein treated her patient deliberately in such a manner as not to give her the slightest hint of what will emerge from the unconscious, and in the process obtained, among other things, the identical scenes with the father.

The import of this letter is clear: Freud still believed, at the end of 1897, that the women had been telling him the truth all along. They *had been* abused. "Why," I asked Anna Freud, "did you omit this letter?" Her answer astonished me. "Because it was wrong. My father later came to believe that the women had been fantasizing, and this letter would only confuse the general public." When she saw my look of astonishment she added, "Just as it has confused you." I have to admit that Anna Freud was far more astute, in many ways, than I was. There was a clarity to her intellect that I could admire without reservation. Nor was she in any way a devious person. She was honest, and straightforward, but there were also things she could not absorb.

I wanted to absent myself, in some personal way, from the significance of what we were finding. I wanted Anna Freud to look at the evidence entirely on its own merits, with no reference to me, to her, or even to the fact that Freud was her father. This was impossible. From her response to me I began to get the first inkling of what was later to become much more explicit and personally virulent.

"Why, Dr. Masson, are you so fascinated by this episode?" Anna Freud asked me. I did not know, but whatever the answer, it was irrelevant to the importance of this momentous historical event. I didn't want to plead with Anna Freud to be empathic with victims of child abuse, it seemed so crude. I wanted her to see that this whole "episode," as she insisted on calling it (as if it were an unfortunate episode, one best forgotten, which is probably what she thought), was central to her father's development of psychoanalysis. However one judged it, its importance could not be denied. One could not pay too much attention to it. Here was one of the central moments in the birth of modern psychotherapy. Freud was confronted with a deep challenge: either he believed the women and turned his considerable intellectual powers onto the issues these hidden truths raised, thereby becoming an ally of women, or he did not believe them, and instead spent his time trying to discover the reasons for their "hysterical mendacity," as he later came to call it, becoming their psychical prosecutor. It was important to decide who was "denying"—society, or the women.

We were at one of our frequent intellectual impasses. "Dr. Masson, my father based his rejection of these women's memories on clinical material. He recanted because he was wrong the first time." (Whenever she used that phrase "my father" I would shudder a bit at its historic magic—knowing, too, that in just a few years, nobody else would ever be able to say that again, but I also learned that it usually prefaced a final opinion and one with which I would not agree.)

I had no choice but to be direct.

"How do you know that, Miss Freud? After all, we have no clinical records demonstrating this to be true. Did he tell you that? Did he ever give you any information on this point? We saw that his reliance on Ellis was unfounded." Silence. Too prolonged for my comfort.

"He told me that he changed his mind based on his clinical work with patients."

"But there is no proof that that is true. There are no case histories. There is no evidence for it in his own clinical notes. In fact, Miss Freud, you must recognize that the very passages from his letters to Fliess that you omitted contain clinical material of the opposite nature to the one you suggest: proof that he was seeing the reality of sexual abuse in his own private practice, *after* he supposedly abandoned the theory. And you omitted those very case histories from the published record yourself. So, in fact, the history of the whole episode has been more obscure than necessary." I took a breath.

"Please, believe me," I continued, "I am not trying to be rude to you, nor to impugn the character of your father. How can we know that anybody else would have behaved any differently? But I am trying to remove this whole question from the personal, from you, from me, from your father. I want to find out what happened then. Please don't analyze my motivation, because I really don't count in this at all; just focus your attention on the material. It is so important. Please."

I saw her look at me with a mixture of admiration and distaste. She was not accustomed to being spoken to in this manner. I knew that. She was looking at me and possibly wondering who I was, what I really wanted, what I intended to do with this new information. She was wondering, even though I begged her not to, about my motivation. She was making a clinical judgment about me. It was her profession to do so. I didn't want that. I wanted to get away from the world of analysis for the time being. I knew that I was taking a risk. She could revoke her permission to publish the letters at any time. But I was not interested in getting into a power struggle with her. I wanted to say something to her that was more direct, more human, less theoretical. It had to do with something essential to the way people grew up in the world, and how they became unbearably sad.

Chapter Nine

DISILLUSION

January 1981

Washington, D.C. The Library of Congress. I was now an officer of the Sigmund Freud Archives, and I was meeting with officials of the library, who had just been informed by Eissler that I was to be his successor as director of the Archives. The Freud collection was one of the library's more important acquisitions, both for the collection's scholarly value and for its commercial worth. But for years now it had also been a source of headaches. The problem was that scholars and university researchers were convinced that there were deep, dark secrets locked away in the vaults of the Archives. Eissler would not permit anybody access to more than a few of the thousands of files in the library. Members of Congress had made inquiries under pressure from their constituents. There were catalogues with intriguing information, but most of the actual material could not be seen for a hundred years. What on earth could there be in the Archives that was so sinister, so dangerous, that it could not be seen for a hundred years?

Almost all of the material in the Freud Archives had been collected by Kurt Eissler. The source for much of it was Anna Freud. Anna Freud had given most of her father's letters to Eissler for the Archives, but she still had hundreds of additional letters of Freud's that were destined eventually for the Freud Archives under the custodianship of the Library of Congress. Now Eissler was passing the responsibility on to me. I was, like everybody else, intensely curious about what was inside some of these files that had been sealed away for one hundred years, and I said so to my hosts at the Library of Congress.

Ron Wilkinson, an employee of the library and a formidable scholar who knows more about this collection than anybody else, told me, "But Jeff, don't you know that you have 'administrative access' to all documents in the Archives? You can simply ask to see anything you want. You just break the seal, that's all."

There was irony in the fact that psychoanalysis, which was dedicated to uncovering, to telling the most difficult truths openly, should have at its very center a cache of documents hidden from public view and even from professional scrutiny, let alone historical investigation. It was a delicious thought, to break the seals on documents that were not to be seen by anybody else for another hundred years. I chose one at random, something that was sealed until the year 2012. The document was dutifully brought to me, and I opened it up. It contained a large map of the world. Various small red pins had been stuck into different cities. It was clear that these represented cities around the world in which there were psychoanalytic institutes. It was Freud's private map of conquest. Like a general, he had stuck little red pins into the cities he had conquered for psychoanalysis. The map was harmless. Why, then, was it sequestered? Was it because Eissler thought it would seem so ridiculous as to throw a bad light on Freud?

I began breaking seals left and right. The official reason given for the existence of seals in the first place was that patients and their identities needed to be concealed (confidentiality), and no doubt in many cases this was true. In some instances, the families of the patients would refuse to donate papers unless such restrictions were imposed. In most cases, however, Eissler himself imposed the restrictions for reasons that were often rather obscure. In my opinion, based on an overview of all the material, a great many of the documents were kept from public view if in the eyes of Eissler or the donor they could prove potentially embarrassing to psychoanalysis.

Otto Fenichel, for example, the great left-wing German psychoanalyst (about whom Russell Jacoby has recently written an excellent book) was represented by a series of secret letters. In these letters he issued instructions that when an analytic institute investigated a senior analyst for any infraction of ethical conduct, this information should not be made public, lest it damage the credibility of psychoanalysis. An analyst's real flaws were to be kept confidential within a small circle of elite analysts, who presumably would let the analyst know they knew about his "weakness"—which would permit them, and only them, to be able to control his behavior. This is, of course, the sort of thing that goes on every day in the United Nations (think of the Waldheim affair), or the United States government, as my friends reminded me. But one does not expect psychoanalysis to be no better than the government. Fenichel felt that knowledge of the true state of affairs would lower the public's idealization of psychoanalysis. It was astonishing that such views could come from one of the great progressive thinkers in the field. Where was Freud's vaunted confidence in the truth? How could it be bad for the public to know the truth? It was only bad for the image of psychoanalysis.

It was my decision, made in conjunction with the librarians at the Library of Congress, that this system should be changed. I was in favor of opening up everything, immediately, as soon as I had the authority to do so. To their credit, they all agreed with me. I told Eissler of my plans.

"I disagree completely. But I am an old man. You will be in charge in a few months. You can then do what you want."

I was thrilled. I loved the idea of opening everything up, of making old and secret documents available to anybody who wished to see them. My only disagreement with the library was that they had some old-fashioned rules to the effect that only scholars were allowed to see the documents. I thought that the definition of "scholar" should be self-chosen, and not conferred by an outside authority. Who was to say who was a scholar and who was not?

The problem had arisen very concretely for me in the case of Paul Roazen. He had written what I regarded as a terrible book about Freud called *Freud and His Followers*. He had for years been trying to get certain letters from Eissler. Eissler refused on the grounds that he did not believe that Roazen was a serious scholar. I did not feel that this was sufficient reason to deny him access to a document he wished to see, as long as that document was readily available to anybody else who wished to see it. Eissler had had very unhappy personal experiences with Roazen, which he described in numerous publications. I also knew Roazen from my days in Toronto and did not trust him. He felt the same way toward us. Still, fair was fair, and my principles were clear. It was a heady moment, to believe that I could and would simply reverse years of subterranean secret dealings. A new wind would blow in the corridors of the Library of Congress, and most of the people I spoke to there were happy about it.

I was receiving a modest but decent salary of $30,000 from the Freud Archives, and this was to be increased when I moved from Berkeley to London. I was editor-in-chief of an elaborate series of translations of Freud's unpublished letters that were to be published by Harvard University Press in the coming years. My daily life consisted in talking to people around the world who would work on these editions, in finding letters still missing (which involved, to my pleasure, a great deal of travel), in frequent trips to the Library of Congress, almost daily conversations with Kurt Eissler, a large correspondence, and of course my own research. Eissler had also given me several hundred tapes of interviews he had conducted with Freud's patients, including the Wolf Man and many others as well, and had asked me to begin listening to them and deciding which could be transcribed and perhaps eventually published. It was exciting work.

The first correspondence to be edited were the adolescent letters of Freud's boyhood friend, Eduard Silberstein. I had asked Muriel Gardiner to edit the letters. Later that year (1981) she and I stood in the garden of my house at the top of the Berkeley hills, overlooking the whole Bay Area. She had come to Berkeley to work with me and Marianne Loring, on both the German transcript of the letters and the translation and annotation. Harvard University Press had agreed to publish them. I was interested in the work, but I was also beginning to feel that I was turning into something I did not want to be. I was now projects director of the Sigmund Freud Archives, the interim title Eissler chose for me before handing over his position as director of the Archives, and this meant that suddenly there were a great many demands on my time—people called all the time wanting access to documents and information; they wanted to come to Berkeley to discuss various projects; and I was also asked to speak in various analytic institutes.

I still had to finish my teaching duties at the University of Toronto, since I had been on a sabbatical and then on an unpaid leave of absence. I returned to spend my last six months in Toronto. It was a strange feeling to be back in the city where I had been so unhappy, knowing now that it was only for a few more months and then I would leave permanently. Everything now looked transparent, and there was a dreamlike quality to familiar scenes. Had I really been there? Was this where it all happened? I felt like a survivor returning to the scene of some painful experiment. I was still alive, and I was leaving.

When I returned to Berkeley, I plunged into my new life with zest. The phone never stopped ringing. There was a certain pleasure in it, no doubt.

But at the same time I felt there was something unreal about it. I was under some pressure not to reveal my true feelings about psychoanalysis, my doubts, which until recently I had been only too happy to speak about. I was becoming a functionary. I represented something, the Freud Archives. My voice was no longer mine alone. And because the Freud Archives were so closely tied into the Freud family (Anna Freud was the main donor), I might as well have been a spokesman for the family. This went against a certain streak of independence in me. I liked the idea of representing nobody but myself. No affiliation, no ties, no loyalties. I turned to Muriel and asked her, "Can this last? Can I do it?"

"Well, Jeff, I'm really not sure," she said. "I like you and I like what you stand for in psychoanalysis. But I'm in the minority in the world of analysis. You know, I gave up my private practice, too. I have become increasingly skeptical of most of my colleagues. It's the question of their ethics that gets me. I'm not as concerned as you are with historical questions and finding the historical truth. For example, when you came to visit me a few months ago, and you went through all those papers on the Wolf Man, you were terribly excited by evidence you found that the Wolf Man had, in reality, been anally seduced. I don't know why, it just didn't get to me. On the other hand, the way I see most analysts behave in the everyday world simply appalls me. I know it appalls Anna Freud too, because we've talked about it plenty. I can't tell you how many analysts come to me to 'confess' to sleeping with patients. Or the things I hear about from my own patients, most of whom have been sent to me by another analyst who is having trouble with them. I have that kind of reputation. I am discreet, it is true, but sometimes I feel like calling them all up and telling them to find another profession. But then there wouldn't be enough left to form a small institute!" The sun was setting, we could see the whole city below us, and it was one of those quiet moments when petty concerns seem to melt away. Muriel finished her drink, and we both sighed and went back into the house.

There was a difficulty in the expectation that I could or would "represent" the standard psychoanalytic view: I did not know what psychoanalysis stood for. I mean this quite literally. I knew what I *imagined* psychoanalysis stood for: the breaking of taboos; fearless invasion into enemy territory, the enemy being ignorance; "speaking truth to power" as we had said in the sixties; abolition of denial; compassion for the suffering of others, especially for those who suffered in childhood; an uncompromising search for historical truth, no matter where this led; finding the hidden injuries of class, sexism, racism. Such was my understanding of the thrust behind Freud's creation of a new discipline, a truth-seeking instrument.

But clearly the thirteen members of the board of directors of the Sigmund Freud Archives and almost all of the leading analysts from around the world thought otherwise. And while Eissler might have agreed with me privately, publicly he too was committed to all of the face-saving gestures common to a public company.

The idea of this job that I now had, or should I say the ideal of this job, was a great deal more impressive than the reality. I attended one of the International Psychoanalytical Association congresses held in Helsinki, Finland. I loved being in a new foreign country, wandering around the streets, seeing the many cars with their headlights still on, a residue from the long endless dark days of winter when there was hardly ever any light, seeing the magnificent tomatoes in the outdoor markets, talking with women in the streets who were delighted to chat with a foreigner.

But as the representative of the Archives, I was also included in a meeting that did a great deal to open my eyes to the political realities of psychoanalysis. I ought not to have been shocked, but I was. As I've indicated earlier, I knew of the practice of soliciting money from patients, but this was the first time I was directly involved. About twenty analysts, most of them holding some official position or other, had gathered together to discuss the funding of the new chair of psychoanalysis at the Hebrew University in Jerusalem. A million-dollar endowment was needed. How to get it? The solution, said one prominent analyst from Chicago, was easy, and had been used several times. "I would ask each of you to compose a list of your wealthiest patients, with their names, addresses and phone numbers. We will then circulate this list within this group. The next stage is for some of us to contact these people, without, of course, telling them how we have their names, and asking them if they wish to donate money for the chair." This was, by any standard I knew, unethical behavior, but nobody in the room voiced any objection.

When I complained to Eissler, he surprised me by defending the action. "Why should not people who have benefited from analysis use their money to help other people to benefit?" Well, I couldn't see anything wrong with that, provided they thought of the idea on their own. But to have their analyst surreptitiously hand over their name to an outside source was still, in my books, unethical, and it was taking advantage of knowledge (the financial state of the analysand) that was privileged and confidential.

"The whole thing stinks, if you ask me," I told Eissler.

"Well, it's done in all fields, and Freud did it within psychoanalysis as well. You know all the wealthy patients who benefited in the early days of psychoanalysis."

True, I did, but somehow it seemed more "pure" then, when psychoanalysis was still struggling, and when there was not a lot of money in it. Today there seemed something reprehensible in it. I just didn't like it.

If I didn't like the direction that psychoanalysis seemed to be going at a practical level, it was not any better at the theoretical level, either. For example, when I spoke about Sandor Ferenczi, Freud's beloved disciple, in San Francisco at an analytic meeting that was open to the public, my old friend Vic Calef could hardly wait for the question period.

"Never mind why Freud did what he did to Ferenczi. Never mind the question of child sexual abuse. I don't really care about Freud and Ferenczi. I care about you, and I want to know about you. Why did you write this paper? What are you really trying to do?"

It was not an answerable question. Even from the point of view of the classical analyst, which Vic Calef certainly was, the question was out of place. A real answer in this framework, the kind of answer that an analyst would want, could only be found in the course of a psychoanalysis. My "motivation," whatever it was, could not be so simple a matter as to be easily inferred by the audience, or even acknowledged by me. I would argue, naturally, that my motivation was irrelevant to the truth or falsity of the material being discussed. I would not claim that it was without interest, only that it was not practically or even logically answerable within the confines of a lecture. One could, in theory, ask any speaker, "What relationship does this paper have with your earliest conflicts with your mother?" Depending on your position, the question may or may not appear vacuous or fatuous, but it is nevertheless one that belongs to another context.

Nowhere did the disparity between what I ostensibly wanted to do and the response my talks elicited from colleagues become more apparent than in a talk I delivered to the Western New England Psychoanalytic Society in New Haven in 1981 on the documents I had been uncovering in Europe. I was the houseguest of Dick and Lottie Newman, close friends of Anna Freud. Dick was a training analyst. Lottie, whose native language was German, and I were working on the Freud/Fliess letters together, and enjoying ourselves immensely. At one point while we were reading the letters we came to an important passage that had not previously been published about the reality of child sexual abuse. Lottie was taken aback by Freud's acknowledgment of the reality of child abuse, and called Dick in to see the letter for himself. After reading the passage, Dick said that he now understood a story that his friend Heinz Hartmann had told him many years before.

"Hartmann told me how shocked he was to see the response he elicited in a very intelligent woman patient. She had told him about her abuse by her father at a very young age. Hartmann said, 'I explained to her, very patiently, that such memories were nothing more than wish-fulfilling fantasies. The abuse never took place, only the desire for it. She said nothing at the time. The next day she came to her session on time, but instead of lying on the couch she told me, with exquisite politeness that nevertheless felt like disdain, that she would not be returning to analysis. I guess the truth of what I told her was simply too much for her to bear. It was too great a narcissistic wound to be told that her most important memory was only a fantasy.' I agreed with him, of course. But now, looking at these suppressed letters of Freud, I am beginning to think the patient was right to leave. She knew, better than Hartmann, that she was *not* caught in the grip of a fantasy. Thanks for showing me the passage." I remember this conversation quite clearly for a reason that will soon be apparent.

I had been apprehensive about the talk I was to give. I knew that many analysts connected with Anna Freud and many analysts from the Yale Child Study Center would be there. The community was certainly classically oriented, but it was also a little stodgy, rather conservative, if not in its analytic views then at least in its quiet New England demeanor. A lot of these analysts smoked pipes. All the men would be wearing ties. The women looked demure. I was probably not dressed properly. As we drove over to a university building where the talk was to be given, I told the Newmans about my apprehensions. They were sympathetic, but were sure that the analysts assembled would appreciate, at the very least, the new material I was going to be talking about. Lottie warned me, though, that only one or two of them knew German.

I gave the talk. Afterward, when I asked for questions, there was a deathly silence. I couldn't tell if I had bored them, or if they disagreed, or were angered, or disgusted. Only Peter Gay, eager, I imagine, to be friends with the new director of the Freud Archives, told me how important he thought the new material was. I did not admire his scholarship and knew that his support was disingenuous. Finally people began to speak up, but little of substance was said. Once again, what struck me about the response to the historical material was how visceral it was. I had struck a nerve. One woman rose and made a perfectly legitimate point. "If, as you seem to suggest, Freud made a mistake in dismissing his patients' accounts of child abuse as nothing but fantasy, then psychoanalytic theory based on this change is questionable. Do you, for example, accept the central role of Oedipal fantasies? Do you even think Freud had anything valuable to say about female sexuality at all? And let me take it even further, do you believe that every analyst who works within the Freudian framework, by *accepting* the reality of Oedipal fantasies, is making a gigantic mistake?"

My answer was to haunt me. "Yes, I do think Freud made a gigantic mistake. Every patient whose memory of abuse was treated as nothing more than wishful thinking will have to be recalled. Excuse the levity, but it reminds me of the Pinto. To answer the rest of your questions, I do not believe that children have Oedipal fantasies about their parents, if by that you mean that girls want to sleep with their fathers. On the contrary, I think parents act upon their children quite literally by hurting them, and in the case of fathers, sexually abusing them. Frequently, in fact. As frequently as Freud at first suspected. And, to be very frank, no, I don't think Freud had anything terribly important to say about female sexuality. Freud's views on women were simply prehistoric. In the *New Introductory Lectures on Psycho-Analysis,* written toward the end of his life, in 1933, for example, he writes: 'The fact that women must be regarded as having little sense of justice is no doubt related to the predominance of envy in their mental life. . . . We also regard women as weaker in their social interests and as having less capacity for sublimating their instincts than men . . . On the other hand I cannot help mentioning an impression that we are constantly receiving during analytic practice. A man of about thirty strikes us as a youthful, somewhat unformed individual, whom we expect to make powerful use of the possibilities for development opened up to him by analysis. A woman of the same age, however, often frightens us by her psychical rigidity and unchangeability.' But let me stress that I am speaking entirely personally now. Please take into account that I have very little clinical experience, and I am just giving you my impressions. I think the relevance of these historical documents for the practice of current psychoanalysis will have to be studied by people more competent than me."

After the talk, many of the analysts present returned to the Newmans for an after-dinner informal discussion. I felt completely isolated in my views. I could not persuade any of the people present to see that there was something about child abuse that needed to be taken seriously. In desperation I turned to Dick and said, "Dick, tell them what you told me earlier today, about Heinz Hartmann and the patient who left him because he told her that her memory of sexual abuse was only a fantasy."

"I don't know what you're talking about. I never told you any such story." I stared at him in disbelief. He had just told me the story that same afternoon. I had been deeply impressed by it, and perhaps even more impressed by Dick's sensitivity. Clearly the story had bothered him enough, when he first heard it, to retain it in memory. And now, so he had told me, much to his credit, he was able at long last to understand why. Why in God's name was he denying the story the same day he had told it to me? It simply could not be that he was so intimidated by his colleagues that he couldn't stick by his story. Was I *such* a dangerous person to know?

Or was it that the implications of Freud's change of heart really did threaten psychoanalysis? Was psychoanalysis really such a troubled fortress, such a closed world, that this kind of inquiry was intolerable? Did they really believe the things they were saying, or were they like intellectual lemmings who will not think for themselves, will not question what the analytic church tells them. Just beneath the surface, this quiet, sedate group of analysts, who would give every appearance to the external world of immense calm and assurance, was seething with quite different emotions. They all had horror stories to tell of their colleagues, of the candidates, of the training. I know, I heard them all the time, even in New Haven. But faced with any criticism from the outside, anything that threatened to give the world a glimpse of the reality that was inside, they closed ranks, they buried their differences, and they took on the air of the wise and calm sages their patients mistake them for. I had sensed this all along about psychoanalysis, but now it was getting to the point where it was far more obvious, forcing me, it seemed, to a real choice. I could not hide behind ambivalence any longer; I knew the score, so I either got out or I would become like them.

After returning to Berkeley, I was called by the *New York Times*. They had heard about the paper and the response to it and wanted to send a reporter to Berkeley to talk to me about the issues surrounding it. Ralph Blumenthal came to Berkeley, spent a few days talking with me, left, and wrote a sober and intelligent account, sketchy and somewhat popular, but basically correct. I was completely unprepared for the storm it was to provoke within psychoanalytic circles. To this day I am not entirely certain what it was in the article that so infuriated the analytic community. But there can be no doubt about the severity of the anger, even rage, directed at me. The two-part article was published in the "Science" section of the *Times* on two successive Tuesdays, August 14, and August 21, 1981. I happened to be in England when the first part came out. Anna Freud had seen it and called me. "I am surprised at all the phone calls I have been receiving. I can't see anything so terrible in this article." I was relieved. When I returned to the United States, the second part of the article in the *Times* appeared, and I called Eissler to talk with him about it, but he would not accept my call. I knew that something was happening. Finally Eissler could no longer contain himself. He called me, and he was clearly in a rage. What struck me about the conversation, though, was that the rage was not at my ideas—he already knew them well before this article—but at the response it elicited among analysts: "Every day I get many calls, from all over the world about how awful you are. How awful this article is. How bad it all is for psychoanalysis. Is this the way you repay my kindnesses to you?"

"But Dr. Eissler, these ideas are not new to you. We have talked them over many times, and we have agreed to disagree. I am not responsible for what I found in the Archives."

"Who else, then, is to be held responsible?" It was curious logic, and I happened to know where it came from. Freud had once been asked whether a man could be held responsible for his own dreams. "Who else would you hold responsible?" was his marvelous reply. Devastatingly logical, no doubt. A dream *is* undeniably the creation of the dreamer, no matter how much input there is from the external world. But the documents I found in the Archives were there before I saw them. Freud wrote them, not I. It was most peculiar the way the Emma Eckstein episode kept coming up in conversation, as if I were somehow personally responsible for it. I didn't know Emma Eckstein. She was a patient and later an analyst who figured prominently in the unpublished Freud/Fliess letters and who was mutilated by Fliess in a botched operation that Freud had approved. I didn't know Freud. I wasn't born until two years after Freud died. But the responsibility for this unsavory episode, as it was called, was being laid at my door, merely because I had the temerity to have found those letters, and, worse yet, to have published them.

Eissler's rage knew no bounds. He did not like being harassed by other analysts. "Just today Masud Khan called me from London and asked me to dismiss you from the Archives. The board members, all of them, or at least most of them, are asking for the same." But Eissler also knew how the very people making the complaints and demanding my dismissal for what was, after all, a disagreement about the history of psychoanalysis were guilty of extraordinary breaches of ethical conduct. The written rules seemed to be very different from the unwritten rules. As I mentioned earlier, Masud Khan had boasted quite openly of sleeping with a patient. He was seeing both her and her husband, and was now living with the woman, but continued to see her husband in analysis! Other analysts told the story with undisguised hilarity. I heard it, however, not just from them, but from Khan himself, who seemed delighted at all the attention he was getting for this "peccadillo." Khan told me: "Nobody wants to say anything publicly because I know too much about all of them. If we were all to be honest with each other, that would be the end of British psychoanalysis."

Another powerful analyst in London was notorious for disappearing on drinking binges. He would return after a few weeks and carry on his analytic practice and his official duties. His behavior was tolerated for the very simple reason that he held a position of eminence in the society. While I was in London, a young and very sweet candidate was terminated. I knew her, and she came to me in tears and asked me if there was any way I could possibly intervene with this very man, whom I knew. What happened was that her roommate, who was also a candidate, told her analyst that this woman was so depressed she had discussed suicide. He told the training committee, and she was summarily dismissed, on the grounds that such a person could not properly treat patients. So I did go to see the man who had the "drinking problem." He was adamant. Such a person should not become an analyst. I was sorely tempted to say, "And what about you? Everybody, even your secretaries, knows about your mysterious disappearances. Is that better?" I didn't, though I wish I had, and the young woman was dismissed. I met her recently, years later. She is now married to an analyst, and defends what happened to her. "It was absolutely just," she told me, "analytic purity cannot be compromised."

Now I asked, "But why, Dr. Eissler, why? What did I do that was so horrible? Adopt a position about the historical importance of the seduction theory that is at variance with the accepted version?"

"Yes."

"But the accepted version is shallow. It is based, primarily, on the account given by Ernest Jones in his three-volume biography of Freud, and as you well know, Jones was hardly a profound scholar, or even a profound thinker. He always gave the easiest and most convenient account of things. He didn't like complexity. He was not honest about sources. His German was poor. His memory was failing him, his eyesight was going, why in God's name should we accept as gospel truth his version of events that he knew only from hearsay? You yourself saw that he totally misunderstood the Freud/Fliess letters time after time. I don't mean the ideas either, I mean just the words. He just didn't know what the German meant. So he guessed, and he often guessed wrong. You saw that. So why feel bound to his superficial version of why Freud abandoned the seduction hypothesis, especially when we have Freud's own letters, that tell us a much richer, much more complex and interesting story? Freud never even spoke to Jones about this, so what Jones says is not firsthand. Please tell me what my unpardonable sin is?"

Eissler could not then, nor ever afterward, explain to me or to anyone else what, precisely, I had done to bring down the ire of the entire analytic world on my head. Almost overnight I had become the arch traitor, the Antichrist in the church of psychoanalysis. Had I not had the full confidence of Anna Freud? Did I not stab her in the back? Had I not prepared a "patricidal" edition of the letters—meaning that I had prepared an unexpurgated edition that allowed Freud to speak for himself for the first time? Did I not appear at meetings without a coat and tie, thereby mocking the solemnity of the science to which I had been so mistakenly admitted?

Eissler reluctantly allowed me to be present at the meeting of the board of directors of the Sigmund Freud Archives, the meeting at which I would inevitably be fired. I called Anna Freud in London to tell her what was about to happen. It was a strange, honest conversation.

"Miss Freud, I am sure you have heard that Dr. Eissler is going to fire me from the Archives."

"Yes. And I disagree with him. I did not like that second article in the *New York Times*. And I think you are wrong in your views. But I do not see why you should be so severely punished for holding them. On one point, however, I feel that I was deceived by Dr. Eissler. He never told me that you were going to live in my house. My understanding was that you were to be in charge of the library and of the research, but not actually live in the house." I never did find out why Eissler never explained this to Anna Freud. Perhaps he was being discreet, not wanting to bring up the matter of her death, or perhaps he knew she would not like the idea of my living in the house. Of course, as things turned out, I never did live in the Freud house.

"Did the idea of my living in your house upset you?"

"Frankly, yes it did."

"Why?"

"Because my father would not have wanted it."

"You mean, he would not have liked me?"

"I am not saying that. But he would not have wanted somebody like you living in the house. He would have wanted somebody quiet, modest, unobtrusive. You would have been everywhere, searching for everything, going through boxes, drawers, closets, bringing people in, opening things up. My father would not have wanted this." She was right.

The thirteen members of the board of directors of the Sigmund Freud Archives met in Eissler's apartment in New York on November 14, 1981, just about a year to the day after I had been appointed projects director. I was present to defend myself. It was a strange meeting on many counts. It was immediately apparent to me that nobody except Eissler and I had any interest in the issues we were supposedly there to discuss. Nor did those present express any understanding of the historical dimensions of these issues. Many of them seemed rather vague as to what, precisely, the seduction theory was, and what, precisely, my apostasy was. But they were extremely well informed about the gossip value of all the fuss. They knew that Eissler was receiving many phone calls asking for my dismissal. They knew that Eissler and I had been close friends, and that the meeting would be traumatic for one or both of us. It promised to provide entertainment. They were not disappointed.

Both of us were close to tears. It was not apparent to me what, exactly, was happening, and I wonder if even Eissler knew. I couldn't help feeling that Eissler, while he strongly disagreed with my ideas, was angry at himself for feeling that he had to bow to pressure and let me go. Before the meeting we had a private talk. Even at that last minute, he could tell me what he thought of some of his colleagues. I mentioned a figure of some importance in the world of analysis, and told him I thought he was a phony. Eissler readily agreed. Another person had enormous charm, and when I mentioned this, Eissler added, "Yes, but I can't stand him, can you?" I couldn't. So something of the old liking, the old bond, was still present. But mindful as Eissler was of his friendship for me, he was even more mindful of the position of psychoanalysis. After all, Freud had sacrificed everything, certainly mere friendships, for *die Sache* or "the cause," that is, psychoanalysis. So much depended on appearances. The articles in the *New York Times* made psychoanalysis look bad. The directors were not asking why, or whether there was any truth in my claims, but were simply concerned with appearances.

I felt very much like an executive of Coca-Cola whose sin was to have made statements that indicated that I did not believe it was the healthiest drink in the world. I didn't. But these were grown men, and call it idealism or just plain naivete I was sure they would see that there were issues far more important than the relative standing of psychoanalysis, issues like truth, like the suffering of real people who had been damaged by what had happened to them in the past. I did not like to think that psychoanalysis existed in a world of politics. Surely we would all be able to rise above our own petty concerns and consider the issues impartially. I kept urging the members present to forget about me, and to think about the issues. But time and again they returned to me. "You should not have spoken to the *New York Times*," said one, and another added, "You showed poor judgment." "You should have been more discreet," added another.

But I believe that my supposed personality, my supposed motivation, and my supposed hunger for publicity really had little to do with what was bothering these men. I believe that they could not get over the fact that my picture had appeared in the *New York Times.* An analyst should not allow his picture to appear in the paper. This could disturb patients. (I did not have patients.) "Jeff," said one of the stiffer analysts there, "you well know that the analyst should be anonymous, unknown." Here was a discussion they could all participate in. They liked it. "Ideally," said another, "a patient should not even be able to describe the appearance of his analyst. The room should be so nondescript that it could not be remembered." "The operating room, remember the operating room," I was admonished. Purity, severity, sterility. I did not look sterile, I did not feel sterile. I was not an analyst.

One of the older analysts present said that the article had been a personal embarrassment to him as a psychoanalyst. What did I have to say to that? I had nothing to say to that. I did not want to be discourteous and tell him that I really didn't give a damn what personal discomforts he had suffered on behalf of my views about the seduction theory. It just seemed too petty to take seriously. "Moreover"—he was not finished—"you have abandoned all the major tenets of psychoanalysis." This was probably true, but I was not sure how he knew. "Let me read what you wrote," and he proceeded to read from the newspaper account, which everybody had already read. But then he did a most astonishing thing—he added the following sentence: "And now I no longer believe in repression or the unconscious." This was as if in a meeting of senior Vatican officials, one of the cardinals were to announce that he no longer believed in the Holy Trinity, or even the existence of God. When there was a murmur of disapproval I had to object. "But I never wrote those lines. They are not in the article at all. You have simply invented them."

"Yes," he said, "they are here in black and white."

I leaned over to see what he was quoting, and noted that he had penciled in the lines he was citing. I said so. "You have added those lines. They are not part of the article." Several people then said something about disagreements—this was hardly a disagreement, he had simply lied—and shouldn't we stop arguing and get on with the matter at hand? I was burning with indignation, but I could see that I would get nowhere by dwelling on this man's outrageous conduct.

At one point somebody got up and said he objected strongly to a sentence I had written in the paper I had given at the Western New England Psychoanalytic Society, where I had accused analysis of being sterile. "By implication, you accused analysts of being desiccated. I am an analyst, and I do not consider myself desiccated, in fact—" but before he could finish Eissler could stand it no longer. "Stop it. Of course Masson is right about the aridity of psychoanalysis today." The analysts in the room were so used to agreeing with Eissler automatically that there were murmurs of assent around the room. Eissler went on, "But the point is, who is to blame for this? Masson would blame Freud. That is outrageous." A chorus of "outrageous" came back. Finally Eissler lost his composure and launched into a forty-five-minute passionate denunciation of me—a bizarre combination of personal confession and complaints and disagreement, very emotionally volatile. He was trembling, and I think people were frightened that he would collapse. I certainly was.

"Why did you say that Freud renounced the seduction theory to line his pockets with money?"

"But I never said any such thing."

"Yes you did, I heard it from a patient who was present when you said it."

"But that's crazy, excuse me, that's absurd. All I have ever said is what I have said here tonight and over and over, that I believe Freud was under strong pressure to give up the theory by his colleagues and that he probably felt that had he stayed with the seduction theory he would not have been able to develop a psychiatric practice because none of his colleagues would send him patients."

Another analyst told Eissler that he too, had heard the same accusation, but was not free to declare his source.

"You called Freud a coward," said Eissler.

"No, I did not. I said he had a failure of moral courage."

"That amounts to the same thing."

"Well, but I believe he did. I'm sorry, but I believe it. I could be wrong, but surely I'm entitled to draw my own conclusions."

Someone interjected, "Yes, but not as a representative of the Freud Archives." He had a point. I explained that I did not think that scholarship had any allegiance except to historical truth, and that I could not defend something I did not believe.

Eissler was very agitated, and became more personal now. "Has anybody ever loved you the way I loved you? I never refused you anything."

It was true, but I did not know what it had to do with the accusations being hurled at me. I really did not know how to respond.

"Has anybody ever done as much for you as I have? And is this how you repay me?"

I felt terrible, not because he was right about the accusation, but because he was obviously in emotional pain, and I was the cause of it. I had hurt him, there was no doubt about that. But I was not certain what I had done. When Eissler finished I asked him, "Please tell me why I am being fired from my position. I was a full professor at the University of Toronto when you offered me this job, and I gave up a tenured faculty position to accept it. You and Muriel assured me it was for life. I have a family I must provide for. Had I known this was just a trial period or that I needed to espouse the conventional views I would never have accepted the position."

Eissler was calmer now, and he said that all that was true, and he would now tell me why I had to be fired. I was being fired for three reasons. "The first is the article that appeared in the *New York Times*. The second reason is the Zeplichal incident. Do you remember, Professor Masson? In one of the Silberstein letters, Freud told his friend that he was sending him a book by Zeplichal. I asked you to find out who this person was. You looked it up and said apparently he had written a book on geometry. *But you were wrong, Professor Masson.* The Zeplichal Freud had in mind had written a book on shorthand, *not* geometry." Here he paused to look up at me. Eissler was serious and apparently considered this almost a sin. "The third and last reason you are being fired is that you told Anna Freud that a letter published in German from Freud to Karl Kraus contained nine transcription errors. But in fact *you were wrong.* There were only six errors, not nine." Again, he looked absolutely indignant. Eissler was a quirky man, a strange and finally a lovable man. I did not want to push him. I did not want to hurt him. Something terrible was going on inside him; he was not capable of talking about it, but it was real, and I was the source of his pain. I did feel bad for him. But I could not let the others off so lightly. So I turned to them, and I said, "Well, Dr. Eissler has told you the reasons why I am being fired. I want to ask you, do you all feel so strongly about Zeplichal?" For a moment, they looked confused ("Who??"), then there was murmured assent, "Yes, indeed, you got Zeplichal wrong, terrible, a terrible incident." I turned to one of the men, whom I had liked, and I said directly to him, "Do you honestly believe this is so terrible? Do you really believe that?" He did not hesitate. "Oh yes, it's terrible, absolutely terrible." I realized they would repeat anything Eissler told them, even if they could barely pronounce the words. The only people who refused to participate in this charade were Lottie Newman and Ruth Eissler, Eissler's wife. They were silent. Ruth looked as if she were about to burst into tears. She hugged me when we left.

Still, the atmosphere of unreality, the almost surrealistic quality of the meeting, persisted to the end. I felt very much like Alice in Wonderland. I did not know the rules, and I was playing against enormous odds. Everybody in the room was older than I, and certainly everybody in the room thought of himself as wiser than I. There was some attempt to treat me with fatherly kindness but harshness. "Jeff, you made a mistake. Why not admit it?" But I wasn't sure what they were referring to. Yes, possibly speaking to the *New York Times* had been a political error, a tactical mistake. I could see that, I could concede that. But people just shook their heads. "Why," somebody said, "do you persist in being so obtuse? We don't mean that."

"Well," I answered, "if you mean my views about seduction, of course I do not believe they are written in stone. They are, after all, only hypotheses. Of course I cannot know, for certain, why Freud abandoned the seduction hypothesis. Nobody can. All we could do is speculate, and the more evidence we bring to bear on our speculation, the better. What, in my views, do you object to? Where did I go wrong? What have I ignored? What have I not understood? Have I improperly translated one of Freud's statements? Did I miss something? Is there a historical document I am not aware of? Tell me." They were nothing if not honest, for nobody tried to hide the fact that they were largely ignorant of what my views were. They had not read any of the relevant documents with respect to Freud, so in fact they weren't even very clear about what Freud's views were. What, then, were we arguing about? What was at issue? I was. And I could not change myself. I could only feel sad. Sad that it had come to this. Sad for myself, sad for the sorry state of psychoanalysis, which could not tolerate this kind of healthy active dissent. Sad, too, for the patients of these men. For how could such men listen to women telling them about their sorrows and believe them, when they could not dispassionately consider issues of truth in the history of their own science? I knew I could no longer participate in such a world. When the meeting ended, I was no longer with the Freud Archives. I knew that my fall from grace meant that I would no longer have any function within psychoanalysis. Everything would be removed from me. It was. Eventually I received the following letter from the International Psychoanalytical Association:

Dear Dr. Masson:

Since we have not received payment of your dues for the last three months, I have been charged to write you to say that you are no longer a member of the International Psychoanalytical Association.

Sincerely yours

A similar letter came from the San Francisco Psychoanalytic Society and the Canadian Psychoanalytic Society. It took longer for the Sigmund Freud Copyright to get around to firing me as a director, but that happened eventually too. I was stripped of all rank, like a disgraced soldier. I ought to have felt terrible. Instead I felt free.

Epilogue

King Janaka, the legendary ruler of the Kingdom of Mithila in India, was once conversing on top of a hill overlooking his city with a wise Buddhist monk. The monk said, "King, look down and across the valley. Do you see those flames? Your city burns." Janaka was not perturbed. He watched quietly for a few minutes, then turned to the monk and said these words, which have been handed down for centuries in India as the quintessence of wisdom: "*Mithilāyāṃ pradīptāyāṃ, na me dahyte kiñcana* (In the conflagration of Mithila, nothing of mine is burned)." The story is told to demonstrate detachment, and the transcendence of any sense of ownership. What was truly Janaka's (love, for example) could not be burned. I feel a little bit like Janaka without the wisdom. As I look back on my life so far, as I survey the career I have had, both in the university and in the professional world of psychoanalysis, I see flames, and the consumption of my life's work. My bridges are truly burned. But while I feel a kind of sadness and a nostalgia for what might have been, I cannot truly say that I am sorry for the loss.

Was there anything in my early life that predisposed me to the experiences I had during my analytic training, and specifically, to the illusion/disillusion theme? I think there was. I grew up primarily in Los Angeles (with periods in Hawaii, Arizona, Switzerland, and Uruguay). My family was Jewish, but both my parents were vegetarian and heavily involved in Indian philosophy. Discussion around the dinner table was often of past lives, of ancient manuscripts, of far-off places. To a child—

to me—that world seemed immensely mysterious, and highly desirable. I dreamed of hidden Tibetan monasteries, where disciples of ancient "gurus" learned secret teachings.

We even had such a "guru" living with us, on and off, by the name of Paul Brunton (a kind of earlier version of Joseph Campbell, down to the hidden anti-Semitism and the fascination with conservative, semifascistic views), who told us that he had been a Tibetan in a former incarnation. At twelve it never occurred to me to be skeptical. Besides, it was more interesting than the conversations I heard at the house of my best friend's, Larry, whose father owned a rope factory. Brunton was the author of a series of books with resounding titles: *A Search in Secret India; The Hidden Path; The Hidden Teaching Beyond Yoga; The Wisdom of the Overself*. I read these books when I was young, and practically memorized them. I meditated daily, and was very puritanical. When P.B., as he liked to be called, told me about our former lives together— always, of course, as glamorous historical figures—I was flattered and incapable of skepticism. Other planets (he had been on Venus most recently) were more fun than junior high proms, and far less mundane.

I was raised to be Brunton's disciple, and was very taken with him and his teaching (Vedanta) until I went off to Harvard College at nineteen. There, thanks to the scholarship of Daniel H. H. Ingalls, the professor of Sanskrit, and a class with the literary critic I. A. Richards, I slowly became disillusioned with Paul Brunton and with "spiritual" values altogether. In a way, I learned more about them than was compatible with unquestioning assent. When I was forced to realize that Brunton did not really know Sanskrit, and had only the most popular understanding of the ancient texts, I began my education in disillusion. My own slow and painfully acquired ability to read the primary sources in Sanskrit and Pali was the main source of my growing skepticism, and not just about Brunton. I found it impossible to read the early Sanskrit texts, both Buddhist and

Hindu, without recognizing the many contradictions and the mutual hatreds that permeated the different sects. I liked the language they were written in, and I often marveled at the ingenuity of the ideas, but I slowly began to free myself from a belief that they were "the truth." They were simply one among a myriad of beliefs that owed their power primarily to the fact that people were born into them, or had the ideas drilled into them over and over. No religion, I discovered, could sustain prolonged critical scrutiny if one was not satisfied with reading secondary and tertiary texts. This ability to read a text critically I owe to Ingalls, and to Louis Renou, the great French Indologist with whom I studied in Paris.

Psychoanalysis became another secret doctrine for me, complete with a guru, Freud, and complex tenets that promised entry into a world sealed off from the uninitiated. Just as learning Sanskrit had allowed me a different view of ancient India, so too did learning German permit me to understand Freud in a new light. When I had learned enough German to read Freud in the original, I saw that there was a great deal in Freud's writing with which I profoundly disagreed.

I can remember how impressed I was when I read the last lines of the *Studies on Hysteria*, which Freud wrote in 1895:

> When I have promised my patients help or improvement by means of a cathartic treatment I have often been faced by this objection: "Why, you tell me yourself that my illness is probably connected with my circumstances and the events of my life. You cannot alter these in any way. How do you propose to help me, then?" And I have been able to make this reply: 'No doubt fate would find it easier than I do to relieve you of your illness. But you will be able to convince yourself that much will be gained if we succeed in transforming your hysterical misery into common unhappiness.'

I used to think this one of the great passages in psychotherapy. I still do. But it did not really sink in to me, when I first read it, that Freud really meant what he said: Misery is unbearable, misfortune is bearable. Misery is self-created, the misfortune comes from without. Freud is not offering to help a person by altering external circumstances, but only by getting that person to reconsider his or her life. It is the essence of psychotherapy: Look inward. Freud is asking us to shift the direction of our attention, from the external to the internal.

It did not occur to me, early on, to call into question Freud's stance whereby his own mental health became the yardstick by which he evaluated all other people. Freud began with a worldview that was conservative and somewhat rigid. He was a medical doctor, trained as a neurologist and a psychiatrist, and while he did not accept everything he was taught, he did absorb a great deal. Fundamental to the nineteenth-century medical and psychiatric view was the hierarchical structure of society. Doctors were superior. Patients were inferior. Patients were flawed. Constitution was inalterable and very important. Each time I would find evidence of these views in Freud, I would be shocked, but I managed to overlook them.

From a historical perspective, it is clear that Freud was perpetuating a tradition that did not begin with him. Its basic characteristics were that it was male-oriented, ethnocentric, sexist, and rigidly hierarchical. It is no accident that the sexual abuse of children, and in particular girls, should have been a major stumbling block, and finally the decisive point upon which so many women were justly able to throw up their hands in disgust at psychoanalysis and Freud. It is clear, too, that had Freud stuck with his first discoveries, in which sexual traumas played such a major role, psychoanalysis would never have become the accepted and popular theory that it is today, because its criticism of existing society would have been too profound, the implications too disquieting. By blaming the victim, Freud was able to unburden the society of any need for reform or deep reflection. Ultimately, Freud reaffirmed the male code, and did little to disturb the sleep of the world. That is why psychoanalysis fits so nicely into hierarchical structures: it is immensely popular in universities and medical schools. It posed no essential threat to the existing order. It could be safely absorbed. It was, in essence, harmless. Freud's early thesis, however, that children were horrendously abused, was abhorrent to society because it was threatening, dangerous, and true. It would have required, especially on the part of men, self-scrutiny, for which nobody seems to have had the stomach.

At the personal level the conclusion I drew from my direct experience was that I was incapable of suspending my judgment and taking on accepted wisdom. I did not wish to join a club, and at a very subtle level, that is what psychoanalysis was. I almost wrote "that is what psychoanalysis became," but that would be historically inaccurate. Psychoanalysis always was, from the moment Freud found disciples, a semisecret society. This secrecy has never disappeared. Secrecy is always accompanied by a sense of entitlement. Built into the very structure of psychoanalysis is a deep division between patient and doctor, between analyst and analysand. One is sick, one is healthy. One is wise, the other is ignorant. One is enlightened, the other is benighted. One is suffering, the other is calm. In theory, every analyst has had his blind spots and his neuroses analyzed away, placing him *au-dessus de la mêlée*. But this rarely, of course, corresponds to reality. What we really have is a case of the blind leading the blind.

Once you begin to claim vision when you see nothing, you have entered the realm of fraudulence. A psychoanalyst makes, ultimately, the same claims as a religious leader, and they are equally false. In my experience, psychoanalysis demanded loyalty that could not be questioned, the blind acceptance of unexamined "wisdom." It is characteristic of religious orders to seek obedience without skepticism, but it spells the death of intellectual enquiry. All variants of "because I say so," or because the Koran says so, or the Bible says so, or the Upanishads say so, or Freud says so, or Marx says so, are simply different means of stifling intellectual dissent. In the end they cannot satisfy the inquisitive mind or still the doubts that naturally arise when such a mind is confronted with authoritative statements about human behavior.

One of the more interesting objections to my work goes like this: "What you say about psychoanalysis, what you point out, the jockeying for power, the struggle for position, the disregard for the deeper rights of patients, the financial underpinnings, all of these exist in one form or another in every profession I have ever been associated with. Why did you think psychoanalysis would be any different? Why did you believe that somehow psychoanalysis would be exempt from these larger laws of nature, the struggle for existence? Were you just naive, or were you inexperienced, or were you ignorant, or just a slow learner?" In a way this is a fair point. I have had enough of a brush with the medical profession to know that what I describe for psychoanalysis exists in precisely the same form in medicine. I know enough about the academic life to recognize that the university is no haven from the kinds of ills I have described. I have even seen a little bit of the legal profession and acknowledge that the same power struggles go on there. But there is one difference. None of these professions, with the possible exception of the university, makes the kinds of claims that psychoanalysis does. No other profession arrogates to itself the same purity of purpose, no other profession claims quite so passionately the moral high ground for its own.

Objectivity and autonomy seemed to me hard to acquire after an analytic training. As I look back on my training, I can see that much of it was an indoctrination process, a means of socializing me in a certain direction; it was partly intellectual, partly political and even to some extent had to do with class. The guild mattered more than anything else. If this process was successful, it became almost impossible to question any of the major ideas within the parent organization. I don't mean, of course, that psychoanalysts do not question Freud or Freudian teachings. Of course they do, all the time. But they do it within a certain framework and within the guild. Stepping outside of that framework, being willing to question the very foundations of psychoanalysis, is unthinkable for most analysts. For a while, it was unthinkable for me. Life within that framework was too comfortable; the money was good, there were conferences, and university appointments, and honors and friends. The rewards were abundant. There was even a degree of warmth and security in accepting the "wisdom" handed down over the last hundred years. Outside, to be sure, it was a far colder world. I am not even certain that had I not been forced out, I would ever have had the courage to step outside entirely on my own.

I went into psychoanalysis looking for the truth about myself. I believed that psychoanalysis was the one science dedicated to discovering such truths and that I would be assisted in my quest. To be sure, I was looking for more than an objective, sympathetic listener and fellow investigator of my own past. But I trusted the tenets of psychoanalysis that "insight" would come about through an emotional relationship with another "trained" person, the analyst, who is free from the biases and distorting needs of his own unmastered past, and who would slowly and skillfully point out my "transference" and lead me to freedom from illusions. I got the emotional intensity I was looking for, and a great deal more into the bargain, which psychoanalysis in its hypocrisy does not tell you about. In a sense, psychoanalysis is undone by the apparent modesty of its claims. Like the sorcerer's apprentice, it unwittingly unleashes forces it can no longer control. Psychoanalysis, despite its dictate, cannot transform the tumultuous storm of human relationships into the artificial calm of a therapeutic alliance. Neither Schiffer's faults, nor my own, were to blame. What I was searching for, and what psychoanalysis promises, cannot in fact, be given by another person, cannot be found in a theory, or a profession, no matter how well-meaning. It is only, I am convinced, to be had, or not had, through living. There are no experts in loving, no scholars of living, no doctors of the human emotions and no gurus of the soul. But we need not be alone; friendship is a precious gift, and all that we need do to see is remove the blinders.